Islands of the Arctic

Over 200 stunning images from the Canadian Arctic Archipelago, Greenland, Svalbard and the Russian Arctic illustrate the beauty of one of Earth's last wilderness areas. *Islands of the Arctic* describes the evolution of the physical landscape through its 4000 million year history, and the processes that continue to affect the region today. Moving from the formation and influence of glaciers to the history of human exploration in an unforgiving environment and from pingos to musk oxen, this book is an excellent introduction for the Arctic enthusiast and a mine of information for the seasoned traveller.

JULIAN DOWDESWELL is Director of the Scott Polar Research Institute and Professor of Physical Geography at the University of Cambridge, England. He is also a Fellow of Jesus College, Cambridge. His research into the dynamics of glaciers and sedimentation in glacier-influenced marine environments has taken him to a number of areas of the Arctic including Svalbard, Russian Franz Josef Land and Severnaya Zemlya, Iceland, East Greenland and Baffin, Devon and Ellesmere Islands in Arctic Canada. He has also worked on board research ships in the Norwegian–Greenland Sea and in the fjords and on the continental shelves of Svalbard and Greenland. He was recently awarded the Polar Medal by Her Majesty the Queen for his contributions to the study of glacier geophysics and has also received the Gill Memorial Award from the Royal Geographical Society.

MICHAEL HAMBREY is Director of the Centre for Glaciology and Professor of Glaciology at the University of Wales, Aberystwyth. He began his academic career investigating glaciers in Norway, the Swiss Alps and the Canadian Arctic, and has gone on to spend seventeen field seasons in Svalbard and Greenland (as well as a further eight in Antarctica), undertaking geological and glaciological research. In 1989, he was awarded the Polar Medal by Her Majesty the Queen for 'outstanding service on various expeditions to the Arctic and Antarctic' and the Earth Science Editors' Outstanding Publication Award in 1995 for his book, *Glaciers*, published by Cambridge University Press.

Islands of the Arctic

Julian Dowdeswell and Michael Hambrey

CAMBRIDGE
UNIVERSITY PRESS

Photo opposite: A shaft of sunlight on the
iceberg-infested waters of Kongsfjorden in
northwest Spitsbergen epitomises
summertime in the maritime High Arctic,
and provides a silhouette for the three
geologists standing on the shore.

PUBLISHED BY THE PRESS SYNDICATE OF THE UNIVERSITY OF CAMBRIDGE
The Pitt Building, Trumpington Street, Cambridge, United Kingdom

CAMBRIDGE UNIVERSITY PRESS
The Edinburgh Building, Cambridge CB2 2RU, UK
40 West 20th Street, New York, NY 10011-4211, USA
477 Williamstown Road, Port Melbourne, VIC 3207, Australia
Ruiz de Alarcón 13, 28014 Madrid, Spain
Dock House, The Waterfront, Cape Town 8001, South Africa

http://www.cambridge.org

First published 2002

Printed in Italy by G. Canale & C. S.p.A

Typeface 10.5/15 Proforma Book   *System* QuarkXPress™   [SE]

A catalogue record for this book is available from the British Library

Library of Congress Cataloguing in Publication Data

Dowdeswell, J. A.
Islands of the arctic / Julian Dowdeswell & Michael Hambrey.
    p. cm.
Includes bibliographical references (p.).
ISBN 0 521 81333 6
1. Islands of the Arctic.   I. Hambrey, M. J.   II. Title.

G615 .D68 2002   919.8–dc21   2002073769

ISBN 0 521 81333 6 hardback

To our families:

Evelyn, Vicky and Adam Dowdeswell
Robert and Joan Dowdeswell
Frederick and Alison Hambrey
Nicholas and Ruth Hambrey

A young explorer takes his first
snowmobile trip into one of the valleys
near Longyearbyen in Spitsbergen.

Contents

The fjords of East Greenland often have
fine sunny weather during summer, and as
a result the region has earned the
nickname 'Arctic Riviera'. A geologist gazes
out across Antarctic Sund as an iceberg,
derived from the Greenland Ice Sheet,
floats by.

Preface

The Arctic conjures up images of winter darkness, the Midnight Sun in summer, desolate expanses of snow and ice, and a rich and varied flora and fauna. It is perhaps epitomised most of all by the polar bear. Sadly, some parts of this complex region have been degraded by Western civilisation, through mineral exploitation, hunting and the activities of the super-powers during the Cold War years. Yet, despite these activities, very large areas of the Arctic, and especially the islands, remain in their pristine state.

As scientists, we have attempted to understand landscape evolution and the underlying geology of the Arctic islands. We have been particularly concerned with the processes associated with glaciers, and how we might obtain from them information about climatic and environmental change in a global context. Between us, we have spent over forty seasons undertaking fieldwork in the Canadian Arctic, Greenland, Svalbard and the Russian islands. Our work has involved camping expeditions to the interior icefields and to coastal areas, in both summer and winter. We have travelled in vessels ranging from nuclear-powered icebreakers to small boats, by helicopter and light aircraft, on snow-scooters, and on foot and ski. Our activities have brought us into close contact with the natural environment and its wildlife, giving us not only a deep sense of appreciation of the beauty of the Arctic islands, but also an awareness of their vulnerability.

Several of the islands are home to the Inuit, who for centuries lived in harmony with nature. Now, growing numbers of visitors to the Arctic islands come for varied reasons: some to exploit the natural resources, many others to experience the wildlife and scenery from the comfort of a cruise ship, some to take part in challenging wilderness activities, and others to undertake scientific and anthropological research. Our aim with this book, therefore, is to help the visitor to learn more about the natural environment of the Arctic islands and, at the same time, through the use of photographs taken largely by ourselves, to acquire a lasting visual impression of the region, even from a brief visit. We also hope that the reader will become better informed about current environmental issues, so that we all may be encouraged to adopt a more sustainable approach to human activity in the Arctic.

The first chapter, *Introduction: the geography of the Arctic islands*, sets the scene by describing briefly the principal topographic and physical features of each major island or archipelago, starting in the Canadian Arctic archipelago and working eastwards through Greenland, Svalbard and the Russian islands. In Chapter 2, *Evolution of the Arctic islands*, we explore how geological processes have shaped the Arctic over nearly 4,000 million years of Earth history, exemplified by the drift of landmasses as they passed through various climatic belts from an equatorial to their current polar position. Chapter 3 examines the weather and climate experienced in the Arctic islands, and looks briefly at some of the unusual atmospheric effects that may be seen. Chapter 4 deals with the features associated with glaciers, those masses of ice which range from tiny patches of perennial ice to the huge Greenland Ice Sheet, and which, perhaps more than any other physical feature of the Arctic, have the potential to affect human activity in other parts of the world through their contribution to rising sea level. Icebergs and sea ice are the subject of Chapter 5: not only do glaciers deliver large quantities of ice to the sea in the form of icebergs, but also the surrounding seas (especially the Arctic Ocean) are characterised by a widely fluctuating cover of sea ice which represents a barometer of environmental change under a warming climate. Frost action (Chapter 6) is associated with permanently frozen ground with a surface layer that thaws out in summer, resulting in difficult construction challenges, as well as a wide range of unusual ground structures. Other aspects of the landscape of the Arctic islands, such as coasts, rivers and lakes, are the subject of Chapter 7. These features offer a range of highly dynamic environments, influenced by many different processes. Within this context, the next chapter (8) explores the relationships between the environment and its often-abundant wildlife, focusing in particular on those species that can be readily observed by the visitor. We then turn our attention to the impact of human activity (Chapter 9), dealing with the indigenous people, the Inuit, looking briefly at the history of exploration and commercial exploitation, and examining some of the environmental impacts of such activity. Finally, in Chapter 10, we look forwards, examining briefly what the future impacts of human activity on the Arctic may be, and exploring how changes taking place here will have an impact on the rest of the World.

This book is, to a large extent, based on our personal experience and observations, but could not have been written without recourse to numerous other publications produced by organisations and individuals. For more detailed information about particular topics, we recommend therefore that the reader consults a selection of the books listed at the end of *Islands of the Arctic*.

JULIAN A. DOWDESWELL
University of Cambridge

MICHAEL J. HAMBREY
University of Wales, Aberystwyth

A year-round resident of the Arctic islands is the Arctic fox. This well-grown cub has started life in its greyish brown summer coat, but will acquire a thick white coat in winter.

Acknowledgements

In undertaking our research in the Arctic we wish to thank our fieldwork colleagues, notably: John Andrews (University of Colorado), Matthew Bennett (University of Greenwich), Nicholas Cox (Natural Environment Research Council), Evelyn Dowdeswell (University of Cambridge), David Drewry (University of Hull), Tim Duggleby (British Army), Jim Dunker (Northern Illinois University), Chris Elliot (British Antarctic Survey), Anders Elverhøi (University of Oslo), Ian Fairchild (Keele University), Neil Glasser (University of Wales, Aberystwyth), Andrey Glazovsky (Russian Academy of Sciences, Moscow), Mike Gorman (University of Cambridge), Jon Ove Hagen (University of Oslo), Brian Harland (University of Cambridge), Bob Headland (University of Cambridge), Niels Henriksen (Geological Survey of Denmark & Greenland), David Huddart (Liverpool John Moores University), Andrew Knoll (Harvard University), the late Olav Liestøl (formerly Norsk Polarinstitutt), Yuri Macheret (Russian Academy of Sciences, Moscow), Geoffrey Manby (University of Greenwich), Heinz Miller (Alfred Wegener Institut, Bremerhaven), Andrew Moncrieff (formerly University of Cambridge), the late Fritz Müller (Swiss Federal Institute of Technology), Tavi Murray (University of Leeds), Clive Pickton (formerly University of Cambridge), Martin Sharp (University of Alberta), James Syvitski (University of Colorado), Martyn Tranter (University of Bristol), Paul Waddams (formerly University of Cambridge), and Robert Whittington (University of Wales, Aberystwyth).

We thank the following organisations for supporting our research either financially or logistically: the Cambridge Arctic Shelf Programme, the Geological Survey of Greenland, the UK Natural Environment Research Council, The Royal Society, the Polar Continental Shelf Project, the British Antarctic Survey, Norsk Polarinstitutt, the Alfred Wegener Institut in Bremerhaven, the Geological Survey of Canada, the Russian Academy of Sciences in Moscow, and the Arctic and Antarctic Research Institute in St Petersburg.

The work has been facilitated by the support of the institutions in which we have worked: the Scott Polar Research Institute and the departments of Earth Sciences and Geography at the University of Cambridge; the University of Bristol; the Swiss Federal Institute of Technology (ETH), Zürich; the Institute of Arctic and Alpine Research in the University of Colorado; the University of Wales, Aberystwyth; and Liverpool John Moores University. Drew Ellis and Jonathan Tooby produced the maps and diagrams and assisted with photo-reproduction. Stephen Coulson and Robert Headland kindly commented on chapters of the book.

Finally, we thank the following for use of photographs: Linda Aasum, David Burkitt, Neil Glasser, Jehu Karhu, Matti Saarnisto, and Atsumu Ohmura for providing several images from the collection of the late Fritz Müller held at ETH Zurich.

A solitary musk ox on the island of Ymer Ø,
East Greenland.

Chapter 1

Introduction: the geography of the Arctic islands

What is the Arctic?

The Arctic, in many people's eyes, is a region of ice and snow, the home of the polar bear. However, this view is only a partial representation of the character of the Arctic, which ranges from ice-covered mountain ranges to vast, monotonous swampy plains of shrubby vegetation that are breeding grounds for mosquitoes. Offshore, the ocean, the habitat of many rare animal species, remains ice-covered for much of the year. In other words, the Arctic is a region dominated by a deep ocean centred on the North Pole that is surrounded by the continental masses of North America and Eurasia and their associated islands, whilst the northern extremities of the Atlantic and Pacific oceans push northwards to the limits of the land.

The Arctic as a whole is difficult to define – it depends on the perspective that is taken. If the Arctic is regarded as the land of the Midnight Sun (in summer), then the southern limit is the Arctic Circle at a latitude of 66°33′ N. If it is thought of as land where ground remains permanently frozen and bears the characteristic tundra vegetation, then the southern boundary is very different – it would exclude Scandinavia but, perhaps surprisingly, include parts of China. Again, the southern limit of sea ice could be taken, but this, too, shows wide latitudinal variations. Climate alone provides a poor basis for defining the Arctic, since it ranges from cool, temperate, wet conditions on the north Norwegian coast, to the hot, dry summers and very cold winters of central Siberia. So how do *we* define the Arctic? In essence, we take a flexible approach by considering the Arctic as having a group of attributes that are concerned with climate, the presence of ice and snow, a

Glaciers meet the sea in the south Spitsbergen fjord of Hornsund.

Map of the Arctic with
the North Pole at its centre.
The island archipelagos and seas
of the Arctic are shown. The form of the
Earth's surface is illustrated in this
illuminated view of the topography of the
major mountains and ice masses.

unique fauna and flora adapted to harsh conditions, sparseness of population and remoteness – and not having precise boundaries.

In dealing principally with the Arctic islands, rather than the whole Arctic region, we still need to take on board the same attributes, which means that some of our selected islands, notably Greenland and Baffin Island, extend south of the Arctic Circle. The Arctic islands epitomise the varied nature of the region and include the Queen Elizabeth Islands of northern Canada (including Ellesmere and Devon islands); Greenland (the world's largest island); Svalbard and Franz Josef Land, which are archipelagos at the northern edge of the European continental shelf; and Novaya Zemlya, Severnaya Zemlya, the New Siberian Islands and Wrangel Island to the north of the Siberian mainland. Each of these island groups features in this book.

Indigenous peoples like the Inuit have lived in the Arctic islands for millennia. Over the last few centuries, however, the Arctic has been explored and exploited by people from the South. In the marine realm, the whale population was severely reduced, followed by walrus and seals. On land, trapping for furs seriously undermined populations of wolf and Arctic fox. Even as recently as the early 1970s, the polar bear, sometimes known as 'King of the Arctic', was taken to the verge of extinction by helicopter-borne marksmen, until brought under the protection of an international agreement. Industrial demands have led to exploitation of mineral wealth, from metals to coal and, most recently, oil. Early operations have left scars on the landscape that will take centuries to heal but today, fortunately, most operations are undertaken in an environmentally responsible manner. In the post-war period we have seen the despoilation of parts of the Arctic through militarisation during the Cold War, manifested most profoundly by the decaying hulks of nuclear installations and vessels in the Russian north – a problem that remains with us to this day, and means that parts of one of the largest islands in the Eurasian Arctic, Novaya Zemlya, remain effectively out of bounds.

Lastly, the Arctic faces a more insidious threat, unconnected with any activity in the region itself: global warming. It is known, for example, that the Arctic sea ice is shrinking in area and thinning, whilst most glaciers are thinning and receding, contributing to sea-level rise. An ozone hole has developed, as in the Antarctic, and climate certainly seems to be getting warmer. Indeed, it is thought by many scientists that the global effects of climatic change will be felt most strongly in the Arctic, with profound effects on the ecological balance.

Nowadays it is widely recognised, at least in the West, that the Arctic should no

longer be exploited so uncaringly. Thankfully, much of the region is still true wilderness, and deserves to remain so. More and more people are able to visit the Arctic. They include scientists attempting to learn what makes the Arctic environment function, adventurers who seek new challenges in a difficult environment, and tourists who wish to absorb the beauty and see exotic animals in their natural habitat. Groups range in size from just a handful of scientists undertaking fieldwork from remote camps to ships carrying several hundred tourists. All will be touched by the experience and, we hope, be convinced of the need to secure a sustainable future for the Arctic.

The scope of this book

Many books and articles have been written, and television programmes made, about the Arctic and its wildlife, but few have covered the physical environment – the physical and biological landscape and its evolution. We hope that this book will serve as a source of information about the landscape, and how it influences, and is influenced by, animals (including humans) and plants. Photographs, selected primarily from the authors' own collections, are intended to convey to the reader the wide range of natural phenomena on the Arctic islands, in the hope that he or she will be better informed about the region. The book is not intended as a regional guide, but rather takes a thematic approach, focusing primarily on the natural processes that have shaped the Arctic.

After setting the scene in this chapter, we describe in Chapter 2 how the Arctic has evolved through nearly 4,000 million years of geological history, documenting its fragmentary evolution, and slow passage through many different climatic and vegetation belts to its present position in the North. Next, in Chapter 3, we look at the nature of Arctic weather and climate, and highlight changes that have taken place in the last 100,000 years or so. Chapter 4 deals with glaciers and ice sheets, past and present, noting how they grow and recede in response to climatic change, and the products they leave behind. In Chapter 5, we turn our attention offshore and focus on floating ice in the form of icebergs and sea ice, how they vary seasonally, and how they may serve as an indicator of environmental change. On land, the all-pervasive presence of frozen ground and vigorous frost activity give rise to a wide range of phenomena that are unique to the Arctic, and form the subject of Chapter 6.

Coasts, rivers and lakes (Chapter 7) are highly dynamic environments in the Arctic, especially where glaciers supply large volumes of water and sediment, and

Four geologists examining an ice-worn rock outcrop on Ossian Sarsfjellet, Northwest Spitsbergen, as small lumps of ice from calving glaciers drift by.

we discuss several of the features that develop in these settings. An account of Arctic plants and animals, and their interactions with the environment, is presented in Chapter 8. Then, in Chapter 9, we examine some of the ways in which humans have adapted to the harshness of the Arctic environment, and also consider the history of exploration and the effects of exploitation and environmental damage in this ecologically sensitive region. Finally, in Chapter 10, we consider measures taken to protect the environment, together with the possible consequences of future climatic change.

The Arctic seas

The Arctic islands form a series of archipelagos between the northern margins of mainland North America, Europe and Asia, and the relatively enclosed waters of the Arctic Ocean surrounding the geographical North Pole. The islands are separated from one another by seaways. Some of these seaways are very large: for

Sea-ice floes stretch away into the distance in this late summer photograph of the very cold East Greenland Current, which flows down the east side of Greenland from the Arctic Ocean. By this time of year, the water is often fairly open, but strong winds can concentrate the ice into lanes of very densely packed ice floes.

A freighter canoe moves through the calm waters of Frobisher Bay, the southernmost of the many fjords on the east side of Baffin Island. Although the land surrounding the fjord contains few glaciers, an iceberg has drifted in from Baffin Bay.

example, the Norwegian–Greenland Sea and Baffin Bay separate Greenland from continental Norway and Svalbard to the east and from Baffin Island to the west, respectively. Lancaster Sound, Viscount Melville Sound and M'Clure Strait form the Northwest Passage, defining the southern boundary of the Queen Elizabeth Islands, and Nares Strait lies between Greenland and Ellesmere Island.

There are also several large seas surrounding the Arctic Ocean, bounded in part by the Arctic island archipelagos. The Barents Sea, at the western edge of which lies the small island of Bear Island (Bjørnøya), is bordered to the north by Svalbard and Franz Josef Land, to the east by Novaya Zemlya, and to the south by continental Norway and Russia. The Kara Sea is located to the east of Novaya Zemlya, and its eastern boundaries are marked by the archipelago of Severnaya Zemlya and the Siberian Taymyr Peninsula. North of the eastern Siberian mainland are the Laptev and East Siberian seas, with the archipelago of the New Siberian Islands (Novosiberskiye Ostrova) between them. Wrangel Island separates the East Siberian and Chukchi seas, the latter located northwest of the Bering Strait between

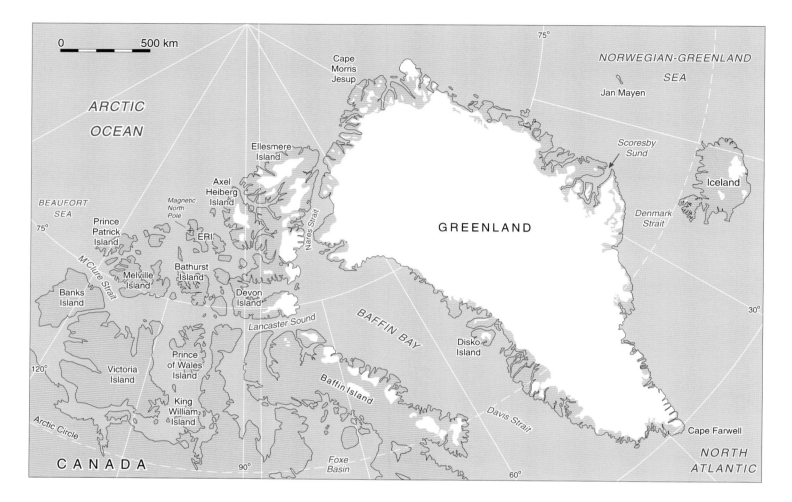

Map of Arctic islands, showing Greenland and the Canadian Arctic. ERI is Ellef Ringmes Island.

the easternmost part of Russia and Alaska. Finally, the Beaufort Sea lies north of Alaska and the northwestern Canadian mainland, and is bounded to the east by Banks Island.

The Canadian Arctic Archipelago

A rough triangle, with apices near the Mackenzie Delta, the northwest corner of Greenland and the northern tip of Labrador, encloses a large number of islands collectively known as the Canadian Arctic Archipelago, which includes the Queen Elizabeth Islands north of the Northwest Passage. These islands range from relatively flat and low-lying with no permanent ice (for example, Victoria and Prince of

Wales islands), to mountainous and heavily glacier-covered (for example, Ellesmere Island).

Baffin Island is by far the biggest landmass in the Canadian Arctic at almost 0.5 million square kilometres, although about a third of its 1,500 km length lies south of the Arctic Circle. Undulating terrain on the west rises eastwards to a high mountainous backbone that exceeds 1,500 metres in several places and has about 37,000 km^2 of glaciers and ice caps. The highest point is 2,057 m. The east coast is rugged and deeply indented by fjords, among the largest of which are Cumberland Sound and Frobisher Bay. Near-vertical rock walls rise above the fjords, terminating abruptly at relatively flat plateaux edges.

To the north of the Northwest Passage, Devon Island, then Ellesmere Island with Axel Heiberg Island to the west, extend beyond latitude 80° N. Each of these islands is largely mountainous and crowned in several places by ice caps and icefields reaching to over 1,500 m in altitude. The 80,000 km^2 of ice on Ellesmere Island represents the largest ice-covered area in the Arctic apart from the Greenland Ice Sheet. Glaciers descend in many places into fjords, calving icebergs into the sea. A summit of 2,606 m, projecting through the ice in northern Ellesmere Island, is the highest point in North America east of the Rocky Mountains. Many smaller

Devon Island, in the Canadian Queen Elizabeth Islands, is made up of an upland plateau dissected by canyons and covered in many places by glaciers and a large ice cap. In this spring scene, snow covers the canyon walls and glaciers fill their floors.

A glacier flows through the mountainous and snow-covered spring landscape of Axel Heiberg Island in the Canadian High Arctic. A prominent meltwater stream, water-filled only in summer, can be seen dissected into the glacier surface.

A glacier-fed river and waterfall cut through granite cliffs in the outer part of Frobisher Bay, southern Baffin Island. The river- and ice-smoothed rock 'pavement' and the rockfall debris above indicate the force of natural processes operating in this harsh but beautiful landscape.

islands contribute to the complexity of the archipelago, but few rise to more than 700 m. Permanent ice is rare or absent in the relatively low-lying islands in the south and west of the archipelago, for example Banks and Victoria islands.

Settlements are few in the Canadian Arctic islands. Ikaluit at the head of Frobisher Bay in southern Baffin Island is the capital of the recently-formed Inuit homeland of Nunavut. The northermost Inuit settlement in the archipelago is Grise Fiord on the southern shore of Ellesmere Island. Resolute Bay on Cornwallis Island is the communications hub for the Queen Elizabeth Islands, whilst the northern tip of Ellesmere Island supports the military base of Alert.

Greenland

Greenland (Kalaallit Nunaat in the Inuit language) has an area of 2,166,086 km^2 and is dominated by one of the two great ice sheets on planet Earth (the other covering Antarctica). It stretches from 60° to 83° N, and its southern tip lies almost in the same latitude as the Shetland Islands, north of the Scottish mainland. In fact, 1.7 million km^2 or over 80% of the island's area is under ice which, in the interior,

A large outlet glacier reaches the sea in northeastern Ellesmere Island, Canada. The crevassed floating tongue of the glacier (to the left) is calving a series of icebergs into the fjord (to the right), which has a thin cover of winter sea ice. At the top of the picture, the rock and talus of the fjord wall can be seen.

The northernmost Inuit settlement of Grise Fiord, Ellesmere Island, in May 1974. The fjord is frozen and the snow cover provides opportunities for children to play with their sledges (Photo: Fritz Müller Collection).

reaches a thickness of over 3000 m. The ice sheet is held back by a rim of coastal mountains, which in places are buried, or alternatively breached, allowing great streams of fast-flowing ice to discharge huge icebergs into the Greenland Sea and Baffin Bay. The land on which the ice sheet rests has been depressed by the weight of the overlying ice, and the bedrock lies below sea level in many areas. If all the ice in Greenland were to melt, global sea level would rise by about 6 m, flooding many of the world's major cities. However, even in a warming world, this would be a very slow process, taking many thousands of years to complete.

The ice-free areas of Greenland feature spectacular mountains, including the highest in the Arctic, Gunnbjorn Fjeld near the east coast, at 3,700 m in height. The coastline is heavily indented, with numerous fjords exceeding 100 km in length, including the world's longest, Scoresby Sund–Nordvestfjord on the east coast (about 300 km long and 13,400 km^2 in area). In Scoresby Sund, 2000 m-high cliffs extend below water level for a further 1,000 m, the product of deep glacial erosion. Glaciers still calve into these fjords, producing huge icebergs that drift out towards the coast and into the surrounding seas. In contrast, sheltered lowland areas support a vigorous tundra vegetation and abundant wildlife. Greenland boasts the

Central East Greenland is characterised by long, deep fjords and channels between coastal islands. Here we see icebergs, derived from the inland ice sheet, drifting through Antarctic Sund, with Ymer Ø in the foreground.

world's largest protected area, the Northeast Greenland National Park, which embraces the entire coastal tract from Jameson Land (latitude 72° N) on the east coast, via the north coast (83° N) to the northeast at Petermann Gletscher, together with much of the interior ice sheet. The park is uninhabited, except for a few research stations, and environmental regulations are strict.

Greenland is a constitutional monarchy, a status introduced by Denmark in 1849. Autonomous local governance or 'home rule' was introduced in 1979. The population of Greenland in 1999 was 56,087, with some 45,000 living in towns. Most of the modern settlements in Greenland are in the southwest, and are associated with fishing, including the capital Nuuk (formerly called Godthåb) with more than 13,000 inhabitants. This was also the area first settled by the Norsemen during the 'Dark Ages', but abandoned when the cooler climate of the 'Little Ice Age' began in the fourteenth century.

Svalbard and Jan Mayen

Svalbard is the Norwegian name, meaning 'cold coast', for the archipelago that lies on the northwestern edge of the European continental shelf (74°–80° N), although it is geographically nearer to Greenland than to mainland Norway. Influenced by the ocean current known as the North Atlantic Drift, the northernmost remnant of the warm Gulf Stream, Svalbard is the warmest and wettest part of the High Arctic. Svalbard includes the island with the more familiar name of Spitsbergen (sometimes misspelt Spitzbergen), which is the largest in the archipelago. Covering 63,000 km^2, Svalbard is about the size of Scotland or the American state of West Virginia. Indeed, with its 60% ice cover, Svalbard in many ways resembles what western Scotland would have looked like about 10,000 years ago. The rugged western coast of Spitsbergen, with glaciers descending from icefields and peaks of 1,000 m, gives way to relatively ice-free dissected tableland in the centre, with further high icy mountains (up to 1,717 m) in the northeast. To the northeast of Spitsbergen lies Nordaustlandet (The North-East Land), an island the size of Wales that is almost totally buried under two ice caps. Various smaller islands complement the two main islands of the archipelago. Of these, Prins Karls Forland off the west coast of Spitsbergen is the most spectacular. More famous is Bear Island (Bjørnøya), the setting for a novel of the same name by Alastair MacLean, which is located about halfway between Svalbard and northernmost mainland Norway. Numerous fjords penetrate into the interior of Spitsbergen and Nordaustlandet

Lush tundra vegetation watered by thawing permafrost typifies valley floors in East Greenland. Cotton grass and moss flank this small stream in Gåseland, inner Vestfjord.

Nordvestfjord, the innermost part of the 300 km-long Scoresby Sund fjord system in East Greenland. Mountains up to about 2,000 m high delineate the fjord, which is up to 1,500 m deep. Many large tabular icebergs are seen in the main fjord, to the right. At the top left of the photograph, the source of the icebergs can be seen: the fast-flowing Daugaard–Jensen Gletscher, which drains ice from the interior of the Greenland Ice Sheet.

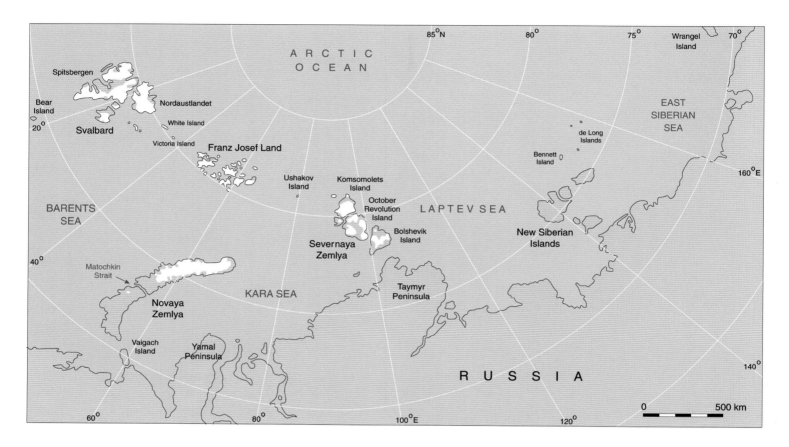

Map of the Eurasian Arctic and its islands and seas.

from the west, whilst the coastal fringes are characterised by narrow, low-lying coastal plains, sometimes known as strandflats, that are home to a rich flora and fauna.

The sovereignty of Svalbard was given to Norway under the Spitsbergen Treaty, formulated in 1920 and implemented from 1925. However, any nation which has signed the Treaty can establish bases there, for example, for the exploitation of mineral resources or for scientific purposes. Both Norwegian and Russian settlements are present; the main Norwegian town is Longyearbyen (population 1,400) and the larger former Soviet settlements are Barentsburg (population 850) and Pyramiden, although since the Cold War ended, the Russians have largely withdrawn from these. Svalbard has no indigenous peoples, nor any reliable record of having had them.

The small island of Jan Mayen, named after the Dutch whaling captain Jan May, is located at 71° N, 8° W in the Norwegian–Greenland Sea. Like Svalbard, it is

Axel Heiberg Island in summer, illustrating the almost bare Expedition River Valley, dominated by the peaks of Little Matterhorn (left) and Bastion (right), viewed from White Glacier.

Evening sunlight breaks through storm clouds above the dissected tableland north of Van Keulenfjorden, south-central Spitsbergen.

Norwegian territory. This isolated island, 53 km long and covering 375 km², lies about 500 km east of Greenland and 550 km northeast of Iceland. The island is the only exposed part of the submarine Mid-Atlantic Ridge north of Iceland. It is capped by an active and largely glacier-covered volcano, the 2,277 m-high Beerenberg, that can be seen from over 60 km away. There is a small Norwegian base and meteorological station close to the eastern shore.

Franz Josef Land

The Russian archipelago of Franz Josef Land (Zemlya Frantsa-Iosifa) lies almost due east of Svalbard, but is much colder. It comprises over fifty small islands at the northern edge of the Barents Sea, and includes the northernmost land in Eurasia, at nearly 82° N. Its dissected tableland resembles that of central Spitsbergen, but is

Early spring aerial view of Bakaninbreen (middle) and Paulabreen (right) in Spitsbergen, showing extensive snow cover over the glaciers and surrounding mountains. Shorefast sea ice in the foreground covers the head of the fjord to a depth of over 1 m.

much more heavily ice-covered. Glacier ice covers about 13,700 km^2 or 85% of the Franz Josef Land archipelago. Glaciers reach the sea along about 60% of the coast, delivering icebergs to the surrounding seas. The highest point in Franz Josef Land is 670 m, but many glaciers and ice caps cover extensive areas at lower elevations.

The largest islands in Franz Josef Land are Vilchek Land in the southeast, and George Land in the west. The completely ice-covered Ushakov Island (81° N, 79.5° E), is almost equidistant between Franz Josef Land and Severnaya Zemlya. Victoria Island (80° N, 37° E), marking the westernmost limit of Russian territory in the Arctic, is located about halfway between Alexandra Land and Svalbard, and is only 60 km east of Norwegian White Island (Kvitøya). The meteorological station that occupied this isolated outpost was closed in 1994. A military airbase on Graham Bell Island, now largely abandoned, provides evidence of the former strategic significance of this High Arctic location. Weather stations still operate on

The same central Spitsbergen panorama in summer – the glacier ice, hidden beneath snow cover during winter, is revealed, together with dark bands of rockfall debris, known as medial moraines. The sea ice has melted from the fjord, and the mountainsides now represent a patchwork of bare rock, scree and remnant snow patches.

A Landsat satellite image of eastern Svalbard, showing the large ice cap of Austfonna on Nordaustlandet (top) and the islands making up Kong Karls Land at the bottom. Between the two lies pack ice at the sea surface. The scale of the image is 185 km across.

A long thin island, Prins Karls Forland, lies off the west coast of Spitsbergen, and in summer is rarely free of cloud. This view, looking north along the spine of the island, shows how the 1,000 m mountains encourage the growth of clouds.

several islands, including Hayes and Rudolf islands, and there is a coastguard base on Alexandra Land. There are no non-governmental settlements or native peoples inhabiting these islands.

Novaya Zemlya

Novaya Zemlya (New Land) consists of two large elongated islands, separated by the narrow Matochkin Strait. It extends north from the Ural Mountains, thus separating European Russia from Asia. The Barents Sea lies to the west, and the Kara Sea to the east. Novaya Zemlya stretches for 960 km from north to south between 71° and 77° N. The islands are mountainous, with summits over 1,000 m. The coastline is heavily indented by fjords. About 24,400 km² or 30% of the two major islands making up Novaya Zemlya is ice-covered, with the vast bulk of this in the northern island. Novaya Zemlya has an infamous reputation as the testing site for many of the former Soviet Union's nuclear weapons, and craters marking the sites of underground explosions can be seen on satellite images of the test area to the north of Matochkin Strait. Submarine reactor cores, with high radiation levels, have also been dumped in the deep fjords on the Kara Sea coast.

Severnaya Zemlya

The archipelago of Severnaya Zemlya (North Land), lying between 78° and 81° N, is made up of four main islands: Komsomolets, Pioneer, October Revolution and Bolshevik islands, together with several smaller ones. The archipelago is bordered by the Kara Sea to the west and the Laptev Sea to the east. Approximately 50% of the 36,800 km^2 archipelago is ice-covered. The largest ice mass is the Academy of Sciences Ice Cap on Komsomolets Island, with an area of 5,575 km^2. Ice shelves (thick floating slabs of glacier ice) are fed from several of the ice caps on Severnaya Zemlya; the best-documented example is the Matusevich Ice Shelf on October Revolution Island which has an area of about 200 km^2.

The Severnaya Zemlya archipelago, which is almost the size of Spitsbergen, was discovered as recently as 1913 and was not explored in any detail until an

The mountainous southern wall of Hornsund in southern Spitsbergen is reflected in the still waters of the adjacent fjord. The inner fjord is covered with a continuous blanket of sea ice, from which the small ice floes seen in the foreground are being produced.

An ice-covered strait separates Hayes Island from a more mountainous landscape to the north, in the central part of the Russian archipelago of Franz Josef Land. The blue colour of icebergs trapped in the winter cover of sea ice can be seen clearly.

Much of the isolated Franz Josef Land archipelago is ice-covered. Ice flows to the coast, where icebergs are calved from ice cliffs about 10–30 m high. This picture shows the coast of George Land in the western part of the archipelago. In winter and spring the sea surface has a continuous cover of smooth sea ice, known as shorefast ice, which is often about 2 m thick.

Fast-flowing outlet glaciers are often heavily crevassed. This glacier, on the eastern margin of a large ice cap on Komsomolets Island in the Russian archipelago of Severnaya Zemlya, produces large numbers of flat-topped or tabular icebergs up to several hundred metres in length. The bergs are embedded in shorefast sea ice for much of the year. In late summer the sea ice breaks up, and the bergs drift into the open sea.

Glaciers flow from left to right between mountains on the east coast of October Revolution Island in Severnaya Zemlya. Rocks fall from the mountainsides to build up ridges or moraines at the glacier edge. Where two glaciers meet, the rock debris merges and is carried forward on the glacier surface, forming ridges known as medial moraines. Spring melting has also formed a small lake in a depression on the glacier surface.

expedition led by the Russian Georgiy Ushakov in the 1930s. The only base of any size in the archipelago is that at Sredniy, or Middle Island, in the Kara Sea immediately west of the main islands. The base has seen recent use as a starting point for several expeditions to the North Pole.

The New Siberian Islands, De Long archipelago and Wrangel Island

The New Siberian Islands (Novosibirskiye Ostrova), between 76°–77° N and 149°–158° E, lie east of the Laptev Sea and northeast of the delta of the great Siberian Lena River. These islands make up the largest archipelago in the Asian Far East sector of the Russian High Arctic. They are low-lying and have no glaciers or ice caps.

To the northeast of the New Siberian Islands are the small islands of the De Long archipelago, located in the shallow Eastern Siberian Sea. Henrietta and Jeanette islands contain small glaciers. Bennett Island, the largest in the group, has an ice cap of 55 km^2 and a maximum elevation of 380 m. These islands remain within the polar pack ice for almost the whole year.

The bleak landscape of Severnaya Zemlya, showing glaciers flowing into a frozen lake. The photograph, taken in April, shows that the archipelago is still in the grip of winter.

One of the early buildings in the predominantly military settlement of Sredniy, in northwestern Severnaya Zemlya. The traditional Russian styling can be seen around the windows. This decorative style provides an unusual contrast to the more functional buildings typical of Russian Arctic settlements.

The easternmost islands in the Russian Arctic are Wrangel Island (Ostrova Vrangelya), at 71° N, 180° E, and the much smaller Herald Island to the northeast. Wrangel Island, whose highest point is 1,097 m, lies closer to Alaska than to the De Long archipelago. It contains no glaciers, but a few semi-permanent snow patches are present. The principal settlement on the island is Ushakovskoye. Wrangel Island is known for the recent discovery of the remains of the last mammoths on Earth, which last roamed the rough summer grasslands of the island about 6,000 years ago in the times of the early Egyptian Pharaohs.

The Neozhdannaya River flows through
the bleak landscape of Wrangel Island,
with snow-covered hills in the
background. (Photo: J. Karhu)

A rifle-bearing visitor (protecting himself against polar bears) surveys the heavily crevassed glaciers of Kronebreen (flowing from right to left) and Conwaybreen (background) in Northwest Spitsbergen. Both these glaciers enter the sea, and in this spring scene the fjord in the foreground is still frozen.

Chapter 2

Evolution of the Arctic islands

The dynamic Earth

The islands of the Arctic have a long geological history. They are the result of processes which have operated over nearly 4,000 million years. We can divide these processes into:

- *tectonic processes*, involving large-scale movements of the Earth's crust, during which rocks are folded and faulted;
- *sedimentary processes*, which involve erosion of pre-existing landscapes and deposition of the resulting detritus in a range of settings to form sedimentary rocks;
- *igneous processes*, involving intrusion of molten rock into the crust from below, and extrusion as lavas to form volcanic rocks at the Earth's surface;
- *metamorphic processes*, whereby the rocks, usually deep in the crust, are altered by heat and pressure.

Whereas most attributes of the Arctic today are related to the polar position of land, this is not so for the rocks. Rather, the islands of the Arctic are built of rocks that once were widely separated and have undergone long journeys to reach their present positions; indeed some originated south of the Equator. It is only in geologically very recent time that the islands of the Arctic have attained their current position. The tectonic processes of continental breakup, continental collision and the formation of new crust on the floor of the oceans, have all contributed to the way in which the Arctic islands have evolved. Underpinning this evolutionary pattern is the all-embracing theory of plate tectonics. Piecing together this history is an

A small motorboat belonging to the 'Sirius Patrol' floats at anchor off the island named Ella Ø, East Greenland. The mountain in the background, rising to 1,800 m, displays a tilted sequence of Precambrian sedimentary rocks, originally deposited on the floor of a shallow sea. These rocks were uplifted to their present position during a mountain-building event around 400 million years ago, but exposed by erosion only in the last 50 million years.

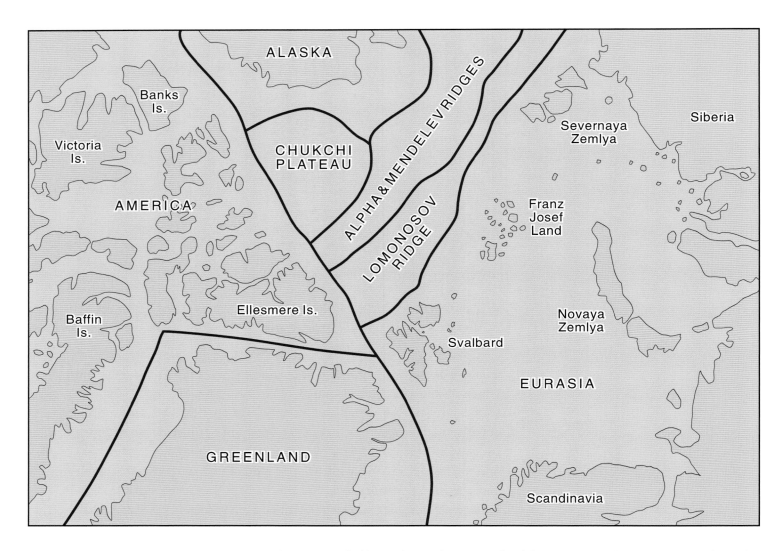

Map of the major geological ('tectonic') plates in the Arctic, shown prior to opening of the present oceans. (Adapted from: D. Worsley & O.J. Aga, 1986, *The Geological History of Svalbard*. Statoil, Stavanger).

important challenge for geologists, rather like trying to solve a jigsaw puzzle with many of the pieces missing.

As soon as new crust is formed, it is subject to weathering and erosion under the influence of climate. Sediment produced from these weathering products is transported by water, wind and ice. It accumulates in valleys and in the sea, in areas that we call 'sedimentary basins'. On land, the sediment provides a home for certain plants, which in turn support animal life. In the sea, marine organisms contribute to the buildup of the pile of sediments. Climate has a strong influence on the nature of these sediments, and experienced geologists can often 'read' the

Snow falls as geologists collect specimens for analysis in the laboratory on a remote 'island of rock', surrounded by ice, known as Tillit Nunatak at the edge of the ice sheet in East Greenland.

past climate (palaeoclimate) after detailed investigations. The rocks of the Arctic islands provide evidence of a complete range of palaeoclimates, from tropical (hot, arid), equatorial (hot, wet) and temperate (cool, moist), to glacial. Thus, a pattern of northward drift of several Arctic islands can be determined, certainly for the past 600 million years. Svalbard and East Greenland are excellent examples of this drift.

Igneous processes are responsible for a range of rock types in the Arctic islands, and they can often be explained in terms of plate-tectonic movements. Continental splitting and separation lead to the formation of new, dense, dark rocks from the mantle, by a process known as sea-floor spreading. This process takes place typically at a submarine ridge, and involves the upward flow of new molten material from deep within the Earth. Existing rocks are pushed aside, so forming new ocean floor. Today, the Mid-Atlantic Ridge is pushing East Greenland and North America away from Europe as new oceanic crust is created. The volcanic activity on the island of Jan Mayen is a part of this continuing process. The same mechanism is widening Baffin Bay and expanding the Arctic Ocean Basin. The rocks that result from this process include basalt and gabbro, which are relatively dense and dark-coloured, full of iron- and magnesium-rich minerals.

Opposite to sea-floor spreading is collision between ocean floor and continental crust, the former (being denser) plunging beneath the latter by a process known as subduction. Subduction is accompanied by violent earthquakes, mountain-building and explosive volcanic activity. Ultimately, an entire ocean may be swallowed up

Million Years	Period		Some key events in Arctic
0			
	Quaternary		Glacial/interglacial cycles
1.8			
	Neogene		Arctic acquires current configuration
23.8			
	Palaeogene		Cool temperate climates Extensive forests ⟶ coal North Atlantic & Arctic Oceans begin opening
65			
	Cretaceous		
142			
	Jurassic		Dinosaurs
206			
	Triassic		
248			Younger
	Permian		sedimentary Tropical climates
290			basins
	Carboniferous		
354			
	Devonian		Old Red Sandstones deserts and rivers
417			
	Silurian		
443			
	Ordovician		Continental collision between Eurasia and mountain building:
495			the Caledonian Orogeny
	Cambrian		
545			
	Pre-Cambrian	Proterozoic Eon	Major ice age Early life forms (algae) preserved
2500			
		Archaean Eon	Oldest rocks, 3800m.y., in Greenland
4600			

A geological timescale showing the major geological periods and key events in the development of the Arctic.

Many transatlantic flights cross Greenland. This view from the flight deck of a Boeing 747, flying at about 10,000 m, illustrates typical rocky terrain of the Laurentian shield, the geological foundation of the Arctic islands. These snow-covered rocks, over 1,000 million years old, are metamorphic rocks, riddled with fractures, picked out by the shadows from a low Sun.

and continent–continent collision may then occur. These processes occur today in the Andes (subduction of sea floor beneath South America) and the Himalayas (collision of the Tibetan and Indian continental blocks). Although the Arctic today is characterised by inactive ('passive') margins, the rocks record an equally violent past. The most dramatic was the Caledonian mountain-building, named after an ancient tribe in Scotland, which resulted in a mountain chain of Himalayan proportions, stretching through Svalbard, East Greenland, the northwest British Isles, western Scandinavia, Newfoundland and the Appalachians. The igneous rocks associated with such events include the relatively less dense granites, intruded into the crust, and extrusive rocks like andesites and rhyolites.

Ancient granite rocks of the Canadian Shield form spectacular cliffs hundreds of metres high on the Meta Incognita ('Almost Unknown') Peninsula on the south side of Frobisher Bay in Baffin Island. The area was first explored by Elizabethan navigator Sir Martin Frobisher in the late sixteenth century, although Inuit had hunted there much earlier.

These mountain-building events were also accompanied by metamorphism and deformation of the rocks. Metamorphism resulted in recrystallisation of both sedimentary and igneous rocks to varying degrees, depending on the heat and pressure involved. The highest temperatures and greatest pressures occurred deep in the crust, typically at depths of 30–40 km. Here, new minerals like the deep red garnet and the shiny, translucent, flake-like mica formed. Typical rock types are gneiss and schist, which show layering defined by specific groups of minerals that formed during recrystallisation. Simultaneously, rocks behaved like a plastic substance and became folded on scales ranging from kilometres to millimetres. At relatively shallow levels in the crust (the top several kilometres), the rocks were more brittle and prone to fracturing. Low-angle faults called thrusts were associated with lateral movements of tens of kilometres, sometimes superimposing older rocks on younger ones. Perhaps most dramatic of all were the sideways displacement of one part of the crust against another, along 'strike-slip faults'. In such cases, movements

The ice-scoured surface in the foreground comprises ancient crystalline gneisses (rocks metamorphosed by strong heat and high pressure) of the Laurentian shield in Gåseland, inner Vestfjorden (Scoresby Sund), East Greenland.

Amongst the strongest evidence of shallow
marine environments during late
Proterozoic time (about 700 million years
ago) is the presence of algal mounds called
stromatolites. These examples have
weathered out beautifully from dolomites
exposed on Ella Ø, East Greenland, and
reach diameters of several metres.

Many parts of the Arctic yield evidence for a late Precambrian glaciation, about 650 million years ago, which some scientists believe was global in extent, an argument that has led to the so-called 'snowball earth' hypothesis. These rocks, called tillites, typically comprise a mixture of all sizes of material from clay to boulders. This example on Ella Ø includes colourful boulders derived from the crystalline shield, which have been transported many kilometres.

may have totalled as much as 2,000 km. Modern strike-slip faults are typified by the well-known active San Andreas Fault in California, which occasionally causes considerable destruction and loss of life, and also changes in the landscape.

The islands of the Arctic do not coincide neatly with geological boundaries. Some, like Svalbard, comprise a number of components referred to as 'teranes' which each have quite different geological histories. By contrast, some now widely separated regions shared a common history until strike-slip faulting or rifting separated them. Despite the complexity of Arctic geology, however, we can identify a number of main stages in the region's evolution.

During the Caledonian mountain-building event, around 420 million years ago, the early sedimentary rocks were folded deep in the Earth's crust, where they can bend more easily. Here a downfold called a syncline is exposed along the 1,200 m-high flanks of Antarctic Sund in East Greenland, as icebergs wallow in the fjord waters beneath.

The geological foundations of the Arctic

The oldest rocks of the Arctic belong to an ancient continent or 'shield' called Laurentia in North America (including Greenland) and Baltica in Europe, two regions which were once joined together. They consist mainly of gneisses, schists and igneous rocks, and represent some of the earliest continental crust found on the planet. Although the greatest extent of these rocks occurs on the Canadian mainland and in Scandinavia, with a small remnant in northwest Scotland, several

The inner fjords of East Greenland reveal spectacular ice-carved rock faces, in which the structure of ancient crystalline rocks is clearly visible. This exposure of Precambrian rock in the inner part of Nordvestfjord, East Greenland, shows large bodies of pink igneous and partially melted material folded in with strongly metamorphosed grey rock.

Arctic islands also possess these rocks, notably Baffin Island, much of Greenland and possibly a part of Svalbard. Indeed, Greenland boasts among the oldest, if not the oldest, rocks on Earth, dating from about 3,800 million years (m.y.) ago. The landscape associated with these ancient rocks often consists of low rocky hills with intervening lakes. The landscape is complicated structurally, and layers and faults in the rocks are picked out by straight valleys and gullies.

At deeper levels, Caledonian mountain-building not only folded the rocks but also metamorphosed them. This close-up view of small folds around a larger fold, in the Hornsund area of Spitsbergen, shows baked layers of sandstone (light) and mudstone (grey).

Early sedimentation, life and climate

The shield areas of today are bordered by belts of less metamorphosed and much younger rocks. These originally formed as sedimentary rocks in late Precambrian and early Palaeozoic time, 1,000–400 million years ago, often in basins that were being stretched. These sediments were laid down mainly in shallow seas, and include extensive carbonate rocks (limestones and dolomites), which possess some of the earliest life forms to develop on Earth. Some of these forms are related to the modern blue-green algae, which disfigure lakes and streams in hot, dry summers. The ancient algae then built up into colonies several metres across, called stromatolites. Spectacular examples of stromatolites can be observed in ice-worn cliffs in East Greenland and northeastern Spitsbergen, whilst in some places individual mounds have weathered out completely, revealing their detailed form. In contrast to these, we find much evidence of an ancient glaciation about 600–700 million

Steeply dipping metamorphic strata of early Palaeozoic age along the flanks of the innermost part of Matusevich Fjord on October Revolution Island in the Severnaya Zemlya archipelago. Icebergs, derived from crevassed glaciers, can be seen embedded within the thin layer of sea ice that covers the fjord.

years ago, that may well have been global in scale. The glacial rocks are called tillites, and are well known from East and North Greenland, and many parts of Svalbard. In addition to these particularly striking rocks, we find extensive sequences of mudstones and sandstones. These sedimentary rocks built up to considerable thicknesses, typically 10–20 km, a figure we can work out because the rocks have been tipped up on end in many places. It is interesting to compare these late Precambrian–early Palaeozoic strata from place to place. In so doing, we find that the rocks in East Greenland are almost identical to those in Eastern Svalbard, even though today an ocean and quite different rocks in western Svalbard separate

In some areas of the Arctic, the end of the Caledonian mountain-building period was marked by intrusion of granite bodies at depth within the crust. Relatively recent erosion has exposed these rocks at the surface. In this view, the granite peak of Newtontoppen (1,717 m), the highest in Svalbard, forms a resistant massif that projects above the surrounding icefield.

the two areas. Thus we can conclude that the rocks in East Greenland and Eastern Svalbard formed in the same sedimentary basin, but that large-scale plate movements have since separated them.

The mountain-building phase

The belts of sedimentary rocks that encircle the ancient crystalline shields, as well as those in North Greenland and the Urals, were next involved in compressional events. These events resulted in the rocks becoming folded, thrust and metamor-

During the Devonian Period the Caledonian mountain chain throughout the Arctic was eroded under a hot desert climate. Coarse material was transported by flash floods to the foot of the slopes to form alluvial fans. These fans typically comprise angular rock fragments and a reddish sandy matrix. This example is a colourful boulder, transported by the modern glacier Fjortende Julibreen in northwest Spitsbergen.

phosed, although in the shallow strata the original characteristics were not everywhere overprinted. Folding was accompanied by mountain-building, a process known as orogeny. The fold belts are typically 100–200 km wide and 1,000–2,000 km long. They not only occur at the edges of the shield areas but also weld them together. Folding took place mainly in the Ordovician and Silurian periods (500–380 m.y. ago). Thus, for much of the succeeding time Europe and North America were united.

In its final stages, mountain-building was accompanied by intrusion of granites as a result of melting of the upper crust. The highest mountain in Svalbard, in fact, is made of such an intrusion. The mountain range was subjected to erosion, and large amounts of red, water-transported sandstones and conglomerates accumulated as large fans at the foot of the ranges. Most of the erosion and deposition took place during the Devonian Period. East Greenland and Svalbard possess these rocks in abundance, and locally preserve early forms of fish as fossils. These rocks are known as the Old Red Sandstone, and similar rocks are forming today in modern tropical deserts that experience flash floods.

Gently dipping strata of Jurassic and Cretaceous age are reflected in the still waters of Van Keulenfjorden, central Spitsbergen. The more resistant beds, recognisable as cliffs, are sandstones, while the intervening scree-draped slopes are made of softer mudstone. Strata like this, where buried beneath the surface, have oil-bearing potential, although exploration to date has failed to yield viable reserves.

Development of younger sedimentary basins

Much of the Arctic now moved into a phase of renewed continental extension. Sedimentary basins developed right across the Canadian Arctic Archipelago, West and East Greenland, Svalbard and Franz Josef Land. Through the Carboniferous, Permian, Triassic, Jurassic and Cretaceous periods (380–65 m.y. ago) sedimentary rocks accumulated sporadically in these basins. Faulting accompanied sedimentation at times, and in places, notably Franz Josef Land, basalt lavas were intruded and extruded. Sediments ranged from marine, where rich shelly faunas accumulated, to

The opening of the present-day North
Atlantic Ocean in Palaeogene time, around
60 million years ago, was preceded by
widespread volcanic activity in East
Greenland and Scotland which at that time
were joined together. This irregular dark
stripe is a basalt dyke cutting vertically
upwards through late Precambrian
sedimentary rocks in Brogetdal, Strindberg
Land, East Greenland.

terrestrial where the first land plants grew and animals such as dinosaurs roamed. Organic matter decayed to produce hydrocarbons which migrated into porous rocks like sandstones to form reservoirs for oil and gas. The search for hydrocarbons has been somewhat fitful in the Arctic islands, although they have been heavily exploited further south (as on the North Slope of Alaska and Northern Siberia).

Opening of the North Atlantic and volcanic outpourings

Towards the end of the Cretaceous Period dramatic changes, already taking place further south, extended northwards into the Arctic – the North Atlantic Ocean was beginning to open, widening the gap between North America and Europe. At approximately the same time Canada was separating from West Greenland as Baffin Bay opened. Prior to these events, East Greenland was connected to Scotland and Northern Ireland, but separation was heralded in both regions by large-scale intrusion of granites (from melted crust) and gabbros and basalts from the mantle. There were also vast outpourings of basalt, forming columnar cliffs and flat-topped plateaux. Igneous events in East Greenland took place over a few million years during the Palaeogene Period (about 52–55 m.y. ago), although sporadic activity as recently as 30 million years ago has also been documented. These events were matched by similar processes in West Greenland and on Baffin Island around 65–55 million years ago. As igneous activity ceased in the continental areas, continued rifting led to the creation of new ocean floor by sea-floor spreading. Baffin Bay and the Norwegian–Greenland Sea, both branches of the North Atlantic Ocean, were initiated in this way. Igneous activity became focused on the central submarine ridges, although these do now project above sea level in Iceland and Jan Mayen. The creation of new ocean floor at these ridges continues to push North America away from Europe to this day at the rate of several millimetres per year.

Elsewhere the Arctic Ocean was developing rapidly into two main basins, the Amerasia and the Eurasia, but in a complex manner associated with the development of several spreading ridges, of which the Nansen–Gakkel Ridge is the most prominent.

Emerging present-day geography and temperate climates

Many parts of the Arctic were uplifted when Palaeogene igneous activity was at its peak, and the island configuration we see today was broadly established at this

Fossil bivalves (*Maletia* sp.) found in Jurassic strata indicate that warm shallow seas prevailed in Th. Thomsen Land, East Greenland about 160 million years ago.

Where molten rocks reach the surface, extensive outpourings of 'flood basalts' can build up to hundreds of metres in thickness. These Palaeogene basalts are dramatically displayed in the fjord walls of Scoresby Sund. The upper one-third of the cliffs in the background comprises numerous lava flows, picked out by a recent snowfall.

time. Western Svalbard experienced a rather localised mountain-building event, deforming older rocks yet again and newer ones for the first time, as the different tectonic plates continued to jostle each other. Extensive forests became established during this relatively warm period, including both broad-leaved and coniferous trees. Beautifully preserved specimens have been found on Axel Heiberg Island and Svalbard; indeed, in the latter area, trees were so prolific that their remains have accumulated into productive coal measures. Clearly the climate must have been radically warmer and wetter than at the present time, even though these continental fragments were already in high latitudes.

Carboniferous (360 m.y. old) strata in Zeppelinfjellet, northwest Spitsbergen were folded during the Palaeogene Period. The low bank of dark grey material is the ice-cored lateral moraine of Austre Brøggerbreen, with the clean glacier surface with a meltstream in the foreground.

The Ice Age in the Arctic

By Neogene time, about 10 million years ago, all the principal elements of Arctic geology were in place, but the region was now to see sharp climatic cooling and the drama of ice ages. Although scientists generally believe that large ice sheets in the Northern Hemisphere only began to grow in late Neogene/early Quaternary time, around 2.5 million years ago, there are clear indications from deep-sea sediment cores that icebergs were sporadically calving at the Greenland coast, and delivering sediment into the adjacent ocean basins, as early as 11 million years ago. How many ice ages affected the Arctic region is unknown. Even the last ice age, known

Sedimentation in the Palaeogene Period in the Arctic was characterised by fluvial deposition under a moist, temperate climate. Extensive forests developed, as manifested by these deciduous leaf fossils in a sandstone boulder on Scott Turnerbreen, central Spitsbergen. Thick organic matter accumulated in swamps, which after prolonged burial was transformed into coal, providing Norway's only coal reserves near the towns of Longyearbyen and Svea.

Winter snow picks out a Palaeogene syncline (downfold) above the glacier Austre Brøggerbreen. Notice how the beds dip in opposite directions at either end of the ridge, thereby defining the syncline. The apex in the ridge coincides with the low saddle. The rocks are Carboniferous in age and are being sampled by the geologist (one of the authors) in the foreground.

as the Weichselian in northwestern Europe and the Wisconsinan in North America, is the subject of considerable debate concerning its extent. Here, it is sufficient to say that many parts of the region were repeatedly buried and modified by ice. The various glaciations, particularly the last, have left behind thick deposits in many places. Modern coastal and river processes are constantly reworking the deposits left by the glaciers.

Chapter 3

Weather, climate and atmospheric effects

The Arctic year

The Arctic islands begin to emerge from the deep cold and darkness of winter in April and May. After days of continuous darkness, the Sun reappears, and rises higher each day until it remains above the horizon for 24 hours. The widespread snow cover sparkles under the Sun and mammals emerge from hibernation and seek out food and opportunities to give birth away from the attentions of predators. Migratory birds begin to return in vast numbers to breed. The snow pack and sea ice begin to melt, although these bright surfaces will remain intact for several weeks yet. Travel, on ski or by snow-scooter, is fast and easy – until the melting snow gathers enough water to generate snow swamps.

Spring merges imperceptibly into summer in June. The snow by now is melting rapidly, and, with the renewal of meltwater flow from glaciers, the large braided river systems fill to the brim and many become impassable. The sea ice breaks up and the broken fragments drift around with numerous passengers – birds, seals and even polar bears. The frozen ground begins to thaw, and extensive areas become waterlogged, hindering travel. The clear days of early spring give way to more cloudy and foggy conditions, as the temperature creeps above zero and the breaking sea ice allows moisture to be released into the Arctic atmosphere. More favoured areas, distant from the coast, may experience warm days of unbroken sunshine, but strong katabatic winds, blowing down the glaciers, can whip up loose material from the ground with little warning. Myriads of colourful but delicate plants burst into flower. Scientists and tourists take advantage of the continuing 24-hour daylight.

A clear spring day on Skobreen, a small valley glacier in central Spitsbergen. The fresh snow sparkles in the sunshine as scientists set up camp.

Thick banks of fog, formed offshore, build up during the summer 'nights' and extend into the inner reaches of fjords. If you are lucky, the fog burns off during the day as in Kongsfjorden, northwest Spitsbergen, depicted here, where a small remnant of fog provides a silhouette for the small icebergs.

Late August in the High Arctic sees the end of the Midnight Sun and the brief autumn season, and the days become increasingly short. The weather often becomes more unsettled; frequent storms occur, bringing mixtures of rain and snow, and fine days become relatively unusual. Migrating birds return with their new offspring to warmer climes, whilst land mammals continue the fattening-up process as best they can. Rivers begin to dry as the freezing nights become longer.

Winter is heralded by more frequent snow storms, increasing darkness – a time for humans to retreat indoors or return south. The sea surface freezes and its

Wind is an important factor in controlling the distribution of snow. At its most extreme, wind carves out clefts between the glaciers and rock walls at high elevations, called wind scoops. This example borders the glacier Haakenbreen in western Spitsbergen.

Coastal fog may persist for many days in maritime areas such as Spitsbergen, but then may lift to form low clouds, at which point drizzle begins to fall. Here a party of geologists has just been dropped off near its field camp in Kongsfjorden from an inflatable boat. Dull, calm days like this are typical of weather conditions during the summer in Spitsbergen.

moisture is once again capped. Snowfall through the winter is usually low, and many areas receive less than a metre in total, allowing grazing animals, such as caribou, musk oxen and reindeer to forage for lichens. In the dark of the long winter, temperatures plummet to −20 °C in maritime Arctic coastal areas, and to as low as −50 °C in the interior of Greenland.

Day and night in the Arctic

The amount of radiation received from the Sun in the polar regions varies more than elsewhere on the globe. This variation is related to the way the Earth orbits the Sun with a tilted axis, producing seasons with extended periods of continuous dark in winter and 24-hour daylight, the Midnight Sun, in spring and summer. At about 80° N, the approximate location of Eureka in Canada, Ny-Ålesund in Svalbard and Sredniy in Severnaya Zemlya, the Sun disappears below the horizon on 22 October, and does not reappear until 20 February, 122 days later. Five degrees of latitude or 555 km further south, Devon Island in Canada, Bear Island in the Barents Sea, and the New Siberian Islands on the margin of the Laptev Sea north of Siberia, lose the Sun for only 92 days. The reappearance of the Sun the following spring is preceded by several weeks of dim twilight under clear skies. Similarly, in summer the Midnight Sun persists for 135 days at 80° N, for 105 days at 75° N, and for only a single day at the Arctic Circle (66° 33′ N). The period over which the Midnight Sun is visible is actually longer than the polar night because the Sun's rays are curved slightly when they pass through the atmosphere, making it possible to see the Sun when in fact it is below the horizon. Anyone staying in the Arctic islands close to the equinoxes will notice how fast the days shorten in autumn and lengthen in spring. Both the native peoples of the North and those operating scientific and technical installations on the Arctic islands celebrate the return of the Sun!

Arctic weather

Winter temperatures in the Arctic islands can reach about −40 °C or below during winter darkness, and rise to about −20 to −30 °C by April, when the Sun returns. Snowmelt usually begins in May, and June is a time of intense melting of snow and ice. In high summer, although the mean temperature in July is no more than a few degrees above zero, temperatures rise into double figures on sunny days, and in windless weather thick jackets can be exchanged briefly for 'T'-shirts. Bright and

With low temperatures and the additional problem of wind-chill, it is essential for Arctic scientists to dress warmly in order to protect themselves from frostbite. In addition, tinted goggles are also required, because the extreme brightness of the snow surface can cause the temporary but painful condition known as 'snow blindness'.

calm summer conditions persist longest in locations furthest from the open ocean, when stable high-pressure cells become established. Examples include the west coast of Ellesmere Island and eastern Axel Heiberg Island in Canada. The east coast of Greenland between about 70° and 76° N has a particularly favourable climate, with summer temperatures rising above 20 °C, and is known as the 'Arctic Riviera'. By contrast, the Russian archipelagos of Franz Josef Land and Novaya Zemlya, in the Barents Sea, are covered with persistent low cloud for most of the late spring and summer. Summer in the Arctic islands is brief, however, and by late August the Sun is low in the sky and the nights are lengthening once more. Winter comes early after the short summer.

As anyone who has lived in the Arctic, or has visited as a scientist or tourist, will know, Arctic weather may change rapidly and dramatically over a few hours

Life in the Arctic benefits from the persistence of the midnight sun in high summer. Here, on the island of Ella Ø in East Greenland, the midnight sun causes the surrounding glacier-capped mountains to be reflected in a small lake.

Probably the best summer weather in the Arctic occurs in East Greenland, an area informally known as the 'Arctic Riviera'. As this view of Brogetdal in Strindberg Land shows, anticyclonic conditions give rise to clear blue skies; these may persist day and night for several weeks. Precipitation is low and glaciers exist only at high elevations. The land becomes bone-dry in spite of the permafrost as summer progresses, except where melting ice and snow water the land.

and even a few minutes. Pilots must monitor changing conditions with particular care to make sure that remote airstrips and rough landing sites remain open. As the wind gathers speed, blowing snow causes 'white-out' conditions, obscuring vision and definition on the ground whilst, ten or twenty metres above, there may be bright sunshine. When blowing snow is accompanied by overcast conditions, travel on the ground by foot, ski or snow-scooter is difficult and often dangerous, because almost nothing of the ground surface can be seen clearly and there is little sense of what is up and what is down. People outside in these 'white-outs' have been known to become disorientated and have walked straight over unseen cliffs in such appalling weather.

Little was known of weather and climate in the Arctic islands until early in the twentieth century. One of the earliest high-latitude weather stations was established in 1875 at Upernavik, at 74° N in West Greenland. A permanent weather station was set up in Svalbard only in 1911, and it was not until after 1945 that continuous meteorological observations began at Canadian Arctic bases such as Resolute Bay and Alert. This expansion of the network of observing stations in the Arctic islands coincided with the increased strategic significance of the region in the Cold War years after World War II. Forecasting the weather in the Arctic became an important part of military information gathering. In fact, during World War II, Germany used submarines to secretly deploy several manned meteorological stations on enemy soil. The remains of two of these stations, set up in isolated locations and resupplied by air or submarine, can still be visited in Krossfjorden, northwest Spitsbergen, and immediately south of Hudson Strait in Arctic Canada. These clandestine stations operated for several years during the war, providing German High Command with weather information that was of considerable importance during the Battle of the Atlantic.

Before the War, weather records had been collected only sporadically by visiting ships and overwintering expeditions to the North. Probably the most

Late August sees the disappearance of the Midnight Sun for the first time (if not otherwise obscured by cloud). Here a Cambridge University motor boat is sailing past the mouth of Isfjorden, as the midnight sun dips behind the mountains to the north.

Late spring clouds at Cape Desire, at the northernmost tip of Novaya Zemlya in the Russian Arctic. The darkest clouds are formed over open water, which gives them their dark appearance relative to those above the lighter sea ice. Clouds like these are known as a 'water sky', and were used by early travellers over sea ice, such as the Russian commemorated in this bleak memorial, as an indicator of open water ahead.

systematic collection of these early records was obtained by the British Admiralty, during a series of voyages to the Canadian Arctic islands between about 1820 and 1850. The seamen of these British Navy ships, which were frozen into the sea ice through one or more years, made meteorological observations of temperature, pressure, wind speed and direction at four-hourly intervals. These measurements were recorded in each ship's log, even during the intense cold of the dark winter months. This information is one of the most complete records of high-latitude climate in what were some of the coldest decades of the past few centuries. Study of these historical records shows that the climate in the Lancaster Sound area was about 2 °C colder in the mid-nineteenth century than that recorded today at the Resolute Bay weather station, on Cornwallis Island north of the sound.

Climate across the Arctic

A large Aeroflot MI-8 helicopter has just landed on the snow surface of an ice cap in the Franz Josef Land archipelago, Russian High Arctic. The helicopter had made three previous attempts to land and resupply the scientists working on the ice cap, but blowing snow had reduced visibility to below the minimum required for a safe landing on the featureless ice-cap surface.

The severity of the climate, that is the average weather over a period of years, varies across the Arctic. In general, it is coldest and driest in the islands that lie furthest from the influence of relatively warm ocean waters. Air masses, usually in the form of warm storms, also penetrate occasionally into the High Arctic from the south, particularly during late winter and spring. In fact, severe late winter and springtime blizzards often bring relatively warmer air, perhaps −5° to −10 °C instead of the more usual −20 to −30 °C that is typical of this time of year, to the Arctic islands. However, to be outside in such blizzard conditions feels extremely cold due to the effects of wind-chill, when the wind removes heat from clothing and skin very rapidly.

A low evening sun illuminates the Isfjord Radio weather station in Spitsbergen against a dark background of fog. This station is located at the southern entrance to the fjord. Spitsbergen has one of the longest running meteorological records in the Arctic, spanning some 90 years.

The warmest Arctic archipelago for its latitude is Svalbard. The mean annual temperature at Longyearbyen (77 °N), the main Norwegian settlement, is −6 °C. Temperatures decrease eastwards across the Eurasian Arctic, as the distance from the relatively warm waters of the Norwegian Current, the northern extremity of the Gulf Stream, increases. The mean annual temperature at Gloermerniy Station in Severnaya Zemlya, 2 degrees further north but 3,500 km to the east of Svalbard, is −16 °C. Similarly, the east coast of Greenland is colder than the west, because the frigid East Greenland Current flows south from the Arctic Ocean, whereas the waters off West Greenland are derived in part from the relatively warmer North

Scientists working in the Arctic islands often augment the network of permanent weather stations by erecting temporary stations in remote locations that record meteorological information automatically. This station was located on the shores of Tinayrebukta in northwest Spitsbergen to help understand the way in which the wind affects tides, currents and iceberg drift in the fjord.

Atlantic. Thus the capital, Nuuk, on the west coast has a mean annual air temperature of −0.8 °C, while that at Ittoqqortoormiit (Scoresby Sund) on the east coast is −6.4 °C. The warmest settlement is Nanortalik near the southern tip with a mean annual temperature of +3.3 °C, a mean monthly temperature for July of 8.5 °C, and for February −6.3 °C. In contrast, the coldest settlement is Qaanaaq (Thule) with a mean annual temperature of −8.9 °C, a July monthly mean of 5.5 °C, and a February mean of −23 °C.

Within the Canadian Arctic islands, too, temperatures are relatively high at −9.5 °C in Ikaluit near the southern tip of the 1,500 km-long Baffin Island. Moving

'Pursuit of science under difficulties'. This plate is reproduced from a nineteenth century book recording one of the British Navy expeditions to the area of the Northwest Passage. During winter the ship was deliberately allowed to freeze into the sea ice, and members of the crew made four-hourly observations of meteorological conditions all through the long polar night despite the possible hazards from polar bears. (From *The eventful voyage of H.M. Discovery Ship 'Resolute' to the Arctic region*, by G.F. McDougall, Longman Brown Green Roberts, 1857).

Sunday, 23*rd*, at 1 A. M., the quartermaster went outside on the floe, to register the thermometers, which were kept in a box secured to a table. In the act of reading off, his attention was attracted

Pursuit of Science under Difficulties.

by a slight noise, and his surprise may be imagined, when, on looking round, he observed a bear within five yards of him. His situation, to say the least of it, was not an enviable one, for one spring of the brute would have put an end to his registering thermometers for ever.

north and west through the Queen Elizabeth Islands north of Lancaster Sound, temperatures decrease along with precipitation. Eureka, at 80° N and on the western side of Ellesmere Island, has a mean annual temperature of −20 °C. Snowfall also decreases westwards, away from the seasonally open water of Baffin Bay and the sea-ice ringed area known as the North Water, which is open year-round. These open-water areas provide sources of moisture to nourish the Greenland Ice Sheet and the large ice caps present along the eastern coasts of Ellesmere and Devon islands. Further west in the Queen Elizabeth Islands, for example on the low-lying Victoria and Banks islands, precipitation is very low, and no glaciers are present.

Climatic change in the recent past

The weather and climate of the Arctic islands has fluctuated dramatically over the last two million years, and is likely to undergo further, and perhaps rapid, change in the foreseeable future. Climatic change in the Arctic islands has been recon-structed through, for example, the study of detailed evidence taken from the annual layers of snow built up in deep ice cores drilled through over 3,000 m of ice at the summit of the Greenland Ice Sheet. These climatic records, produced using the shifting ratio between heavy and light isotopes of oxygen as an indicator sensi-tive to past temperature changes, chronicle environmental changes over at least the past 120,000 years, back to the last interglacial period when the Earth's climate was similar to that of today. A series of climatic shifts between relatively warm and colder conditions is evident from these records, with temperatures falling by 10–15 °C in a few decades and precipitation changing by 50% in only a few years. These are rapid changes in human terms.

About 18,000 years ago, and probably on several previous occasions during the past two million years, most of the Arctic islands were covered by large ice sheets, up to several kilometres in thickness. Temperatures were about 10–15 °C colder than those of today. Svalbard and the Russian Arctic archipelagos, probably as far east as Severnaya Zemlya, were part of a single ice sheet that also filled the Barents Sea and linked together with another major ice dome over Scandinavia and, in turn, with that over the British Isles. The western and northern margins of the ice sheet extended to the edge of the continental shelf, and delivered large quantities of ice-bergs and glacial debris to the Norwegian Sea and the Arctic Ocean. The Canadian Arctic islands were also largely covered by an ice sheet, joined to that over

Lenticular clouds above the icefield of Lomonosovfonna, northeast Spitsbergen, indicate high winds aloft and the imminent arrival of blizzard conditions. Sometimes known as hogsbacks, these clouds form over high ground and remain stationary.

Greenland in the Nares Strait area, and to the Laurentide Ice Sheet that extended southwards to the Great Lakes. It is through the dating of organic material deposited in the adjacent oceans that we can reconstruct the timing and dimensions of these great ice sheets. The ice sheets began to decay from about 13,000 years ago in response to climatic warming and, by about 10,000 years ago, had for the most part retreated close to their modern limits. Glaciers in the deep fjords of East Greenland retreated to their modern limits only 6,000–7,000 years ago.

Over the past ten thousand years, known as the Holocene Epoch, temperatures have been relatively close to those of the Arctic today. Ice cores drilled through up to almost 1,000 m of ice on, for example, Devon Island in Canada and Komsomolets Island in Russian Severnaya Zemlya, show that a gradual cooling has taken place over the past 8,000 years or so, and that the coldest interval during this period was that of the past few hundred years, known as the 'Little Ice Age'. Since about the

A scatter of rusting artefacts marks the site of the German weather station set up during World War II in Signedalen, close to the shore of Krossfjorden in northwest Spitsbergen. Walking in the surrounding hills, the remains of parachutes, coal and even metal cans marked with swastikas can be found hidden among the rocks.

beginning of the twentieth century, however, the climate of the Arctic has warmed by approximately 2 °C, according to meteorological records from Upernavik and Svalbard. This shift is thought by scientists to be related largely to natural variability in the climate system, rather than to human interference.

Human-induced climatic change

Over the past twenty years or so, evidence has begun to accumulate of warming that may be a response to the growing amounts of 'greenhouse' gases, such as carbon dioxide and methane, that are being released into the atmosphere as a result of human activity. A number of years during the 1990s were among the warmest on record. Climatic warming is likely to be more pronounced in the Arctic than at lower latitudes. A major reason for the particular sensitivity of the Arctic to global

Blizzard conditions can occur at any time of the year in the Arctic, especially at elevations above 500 m. Here a geologists' camp at 1,000 m on Wilsonbreen in northeast Spitsbergen has become almost buried by a July blizzard that lasted three days.

Scientists here are emerging from their camp near the head of the glacier Nordenskiöldbreen, Spitsbergen, after a midsummer blizzard, as eddies of wind-blown snow hurtle downglacier in the distance. Although much of the surface was swept clear of snow, the camp has attracted a large drift, even though the tents are widely spaced to avoid burial. Digging out after a blizzard is time-consuming, especially if partial melting of snow has produced an armour-casing of ice.

warming is that sea ice at present reflects much of the received solar radiation. If, however, the sea ice of the Arctic began to thin and melt, then the darker ocean would absorb more radiation, and heating would continue. We will see in Chapter 5 that there is increasing evidence that the Arctic sea ice is doing exactly this. In the context of climatic change, the Arctic is therefore seen by many scientists as an important region for the early detection of global warming.

Beyond the lower atmosphere that controls our day-to-day weather, the ozone layer has also been affected by the waste products of our industrial society. This layer occurs in the stratosphere, at an altitude of about 20 km. It helps to protect us from solar radiation, especially that at ultraviolet wavelengths. The ozone layer has been depleted through the release of chemicals such as the chlorofluorocarbons (CFCs) used for refrigeration during the industrial age. The upper atmosphere in the polar regions becomes particularly depleted of ozone in the spring, when sunlight returns and chemically-active chlorine is produced, and ozone is transformed rapidly into oxygen in the stratosphere. The so-called ozone hole, first detected in the Antarctic, is largest at this time of year, and often extends southwards from the

January

July

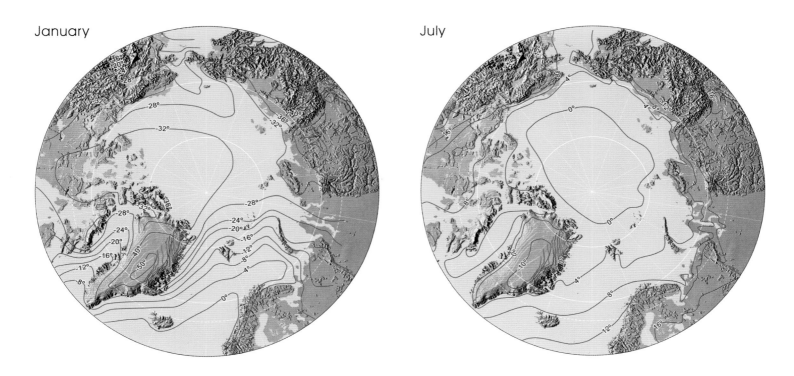

Mean temperatures (°C) experienced in the Arctic during summer and winter. The averages for January and July are shown (Adapted from Orvig, S. (Ed.) 1970. *Climates of the Polar Regions*. Elsevier, Amsterdam.)

North Pole to include the Arctic islands. The effects of this are unseen, but more ultraviolet radiation reaches the ground when the protective ozone layer thins. The concern for humans, and for other animal and plant life, is that this radiation will damage the cells of our bodies, and of our skin in particular, causing an increased incidence of cancers. Clothing and sun-screens provide protection against this radiation hazard for humans, but the long-term risk of any genetic damage to Arctic plants and animals is presently being investigated.

Atmospheric phenomena

The Arctic islands are also a spectacular setting from which to observe several atmospheric effects that are rarely seen at lower latitudes. Blowing snow and ice crystals in the Arctic atmosphere can produce remarkable optical phenomena. Mock suns, haloes and sun pillars are all produced by the refraction of sunlight through the thin curtain of suspended ice crystals. Perhaps the most spectacular optical effect occurs when the Sun appears to be surrounded by a large bright circle

The large and fast-flowing Daugaard–Jensen Gletscher (bottom left) drains almost 10% of the Greenland Ice Sheet into Nordvestfjord, a part of the Scoresby Sund fjord system in East Greenland. The surrounding mountains are about 2,000 m high and the fjord is over 1,000 m deep close to the glacier front. Daugaard–Jensen Gletscher retreated up the 300 km-long fjord to this position only about 6,000–7,000 years ago. Before this, during the last glacial period of cold climatic conditions, the fjord would have been filled by the glacier.

The layers of snow deposited each year
form a continuous record of climate
change when collected as cores and
analysed in scientific research laboratories.
Here, a shallow ice-core is being drilled at
the crest of Austfonna ice cap in Svalbard.

or halo with arcs and mock suns, sometimes known as sun dogs. By contrast, when conditions are calm, the warming action of the Sun on Arctic glaciers and sea ice produces a strong near-surface temperature gradient that refracts light to produce mirages similar to those reported from deserts. We have seen mirages of a tent-camp on the ice from a number of kilometres distant. The camp should have been obscured by intervening undulations in the ice surface, but instead appeared above the horizon as a shimmering and sometimes inverted image.

During the polar night, too, those who remain on the Arctic islands can see remarkable atmospheric phenomena, this time in the upper atmosphere. The Northern Lights, or Aurora Borealis, can light up the winter sky in cloudless conditions. They are caused by the excitation of electrons by the solar wind at heights between about 100 and 300 km on the outer edge of the atmosphere.

Spring in Krossfjorden, northwest Spitsbergen, showing the strong contrast in brightness or reflectivity between open sea water, sea ice and snow-covered land. This difference in reflectivity is a vital control on energy exchange between the atmosphere and the Arctic seas. Where sea ice is present, the energy of the Sun is mainly reflected back into the atmosphere, but in areas of much less reflective water, solar energy acts to warm the sea surface.

A so-called 'Glory' surrounding the shadow of a Twin Otter aircraft as it flies above clouds in the Canadian Arctic. The size of the Glory rings, which form when a shadow is projected onto the clouds below, depends on the size of water droplets or ice crystals in the cloud.

These great curtains of green, blue, purple and sometimes red move rapidly about the sky. The colour of the aurora depends on the type of atom or molecule struck by electrons, as they rain down along the Earth's magnetic field lines; yellow-green colours are produced by oxygen atoms at about 100 km, whereas at 300 km the excitation of oxygen gives a red aurora. A blue aurora is formed from ionised nitrogen. The aurora occurs mainly at high latitudes, where the Earth's magnetic field lines are progressively concentrated with approach to the Magnetic North Pole, which is not a fixed point on the Earth's surface, but presently located north of Ellef Ringnes Island in the Canadian Arctic. The Aurora and the full Moon provide sufficient light for occasional overland travel in the Arctic during the winter months before the Sun reappears.

Optical effects caused by the refraction or bending of sunlight as it passes through falling ice crystals can be very spectacular. An aircraft tailfin at Resolute Bay on Cornwallis Island in Arctic Canada allows a circular halo and associated arcs to be photographed by shading the camera lens from the Sun. The ice crystals are often blown by wind in the lowermost few metres of the atmosphere. Different optical effects are produced through variations in the shape and orientation of the ice crystals in the atmosphere.

The colourful 'curtains' of the Aurora
Borealis or Northern Lights in the sky
above Spitsbergen. (Photo: L. Aasum)

The late evening Sun is reflected on the waters of Kongsfjorden and a partly-frozen river draining Pedersenbreen in northwest Spitsbergen.

Chapter 4

Glaciers and ice sheets

Introduction

Glaciers are among the most dramatic elements of the Arctic, and are responsible for many of the most rapid changes experienced in the Arctic landscape. They influence local climate, control the flow of the largest rivers, and release massive volumes of ice into the sea. Glaciers may provide highways to the interior of mountainous islands or, if heavily crevassed, present an impassable obstacle to travel. They erode the land surface and carry debris to lower elevations, producing a wide range of landscape features. These landforms have counterparts in areas affected by glaciation in the past, such as northern Europe and much of North America. Glaciers are also important to human civilisation as major contributors to global sea-level rise and as long-term monitors of climatic change.

What is glacier ice?

Glaciers are bodies of ice, built up largely from snowfall. The ice is formed in places where some of the snow that falls during winter survives through the following summer. The thickness of the snowpack thus increases year by year and becomes more and more dense as it does so. When all the air passages within the snowpack finally close off, to leave only isolated air bubbles, the snow is said to have changed into glacier ice. The depth at which this happens ranges from only a few metres in places with a relatively maritime climate, in areas like Svalbard and the Russian Arctic islands, to several tens of metres in the very cold, dry areas of central Greenland. The transformation of snow into ice is slower at extremely low temper-

Independent glaciers and outlets of the Greenland Ice Sheet reach the sea around many parts of Greenland. The rugged mountains of Kangerlussuaq Fjord in East Greenland provide the backdrop to icebergs floating in the calm fjord waters.

atures, because the so-called sintering process by which snow crystals grow together proceeds less rapidly.

Glacier size and shape

Glaciers occur in all shapes and sizes. The largest ice masses are known as ice sheets, which by definition exceed 50,000 km² in area. In the colder climates of the past, ice sheets covered up to millions of square kilometres of the Arctic and formed thick domes of ice over the entire landscape. Today, there are only two ice sheets on Earth – those covering Greenland and Antarctica. The Greenland Ice Sheet has an area of 1.7 million km² and is over 3 km thick at its centre, although it is significantly smaller than the 13.7 million km² Antarctic Ice Sheet. The weight of

Part of the Greenland Ice Sheet, pierced by a series of rock outcrops or nunataks. The long shadows accentuate the jagged profile of the mountains, and contrast strongly with the smooth ice surface, which is disturbed in one place by crevasses.

Small icebergs calving from the terminus of the tidewater glacier, Gerard de Geer Gletscher, as it enters Isfjord in East Greenland. This glacier is a major outlet from the Greenland Ice Sheet.

the ice on Greenland depresses the crust beneath by about 1 km, to the extent that bedrock in much of the centre of this island is below sea level, whilst the perimeter is delineated by half-buried mountain ranges. From the centre of the ice sheet, huge fast-flowing outlet glaciers descend through the coastal mountains to the sea, and fast-flowing ice streams form zones of very active ice within the ice sheet. During the last cold phase in Earth's history, about 18,000 years ago, ice sheets also developed over North America, Scandinavia, the British Isles and on Arctic continental shelves such as the Barents Sea.

Ice caps are also domes of ice that largely bury the underlying landscape, but are defined as covering an area of less than 50,000 km². There are many ice caps in the Arctic today. The Austfonna ice cap on the island of Nordaustlandet in Svalbard is the largest in Eurasia at 8,120 km². The 14,000 km² Devon Ice Cap, covering most

Ice caps have smooth, rounded surfaces, and hide most of the underlying bedrock relief. This ice cap covers most of the island of George Land, the largest landmass in Russian Franz Josef Land. The sea surface in the foreground is covered by winter sea ice.

of eastern Devon Island, is an example from the Canadian Arctic. A series of smaller ice caps covers much of the Franz Josef Land and Severnaya Zemlya archipelagos in the Russian Arctic.

Where many mountain peaks break through the ice surface, and the underlying landscape is complex, highland icefields with multiple outlet glaciers are formed. Their surfaces reflect the underlying topography to some extent, but icefields are too thin to bury the mountains completely. Mountains project above the ice as rocky islands called nunataks, a term derived from the Inuit language. Examples of icefields occur on Ellesmere Island, Spitsbergen and Greenland.

Valley glaciers not only flow from the heart of ice sheets, ice caps and icefields, but can also have their own catchment areas. On the smaller scale, mountain hollows are occupied by cirque glaciers, and several of these may merge to form a valley glacier. Representatives of these types occur on almost all of the Arctic islands. Some glaciers flow from confined valleys and spread out as wide lobes onto lowland plains; these are called piedmont glaciers and fine examples occur in the Canadian Arctic on Axel Heiberg and Bylot islands.

Where valley glaciers terminate in the sea, the term tidewater glacier is used, and these are common in many parts of the Arctic. Sometimes an ice cap reaches the coast as a continuous line of ice resting on the sea floor; these are known as ice cliffs or ice walls. Unbroken stretches about 100 km in length are found on the east coast of Nordaustlandet in Svalbard, and on the Prince of Wales Ice Cap in Ellesmere Island. Floating slabs of glacier ice fringing coastal areas, called ice shelves, which have received so much media attention in the Antarctic following wholesale break-up, are rare in the Arctic. Some of the outlet glaciers of the Greenland Ice Sheet have floating tongues up to about 80 km long. They often occur in sheltered fjords, where the sea ice seldom breaks up and protects the ice margin against rapid iceberg calving. Elsewhere in the Arctic, floating ice shelves and tongues fringe much of the northern coast of Ellesmere Island, and are also present on the eastern side of Severnaya Zemlya.

Glaciers may also be classified according to their temperature distribution, or thermal structure. The more familiar alpine glaciers are 'temperate' or 'warm', that is, the ice is at melting point throughout. In contrast, some Arctic glaciers are frozen to the bed, and are known as 'cold' or 'polar'. Many Arctic glaciers are a mixture of cold and warm ice, and are termed 'polythermal'.

Distribution of glaciers on the Arctic islands

The distribution of glaciers, ice caps and ice sheets within the Arctic islands, and indeed globally, is controlled by climate in the form of temperature and snow precipitation. Where summer temperatures induce much melting, or where there is very little snowfall, glaciers will not form even in very cold conditions. For example, it is too dry and too warm in summer for glaciers to build up on the low-lying New Siberian Islands of the far-eastern Arctic, or on Banks and neighbouring islands in the Canadian Arctic. Similarly, large ice caps and glaciers are present on the eastern sides of Baffin, Devon and Ellesmere islands in the Canadian Arctic, close to the moisture source provided by Baffin Bay, but comparatively fewer glaciers are present on the western sides of these islands. We now have a reasonably accurate assessment of the extent of glaciers in the Arctic, based largely on aerial photographs and satellite imagery.

In the Canadian Arctic, Baffin Island has two main ice caps, the Barnes and the

An upland icefield, the Agassiz Ice Cap, on
Ellesmere Island in Arctic Canada. Rock
outcrops or nunataks, highlighted by the
low angle of early spring sunshine, break
the otherwise smooth ice surface.

Isolated peaks known as Alfabet
Nunatakker at the edge of the Greenland
Ice Sheet in East Greenland, viewed from
Tillit Nunatak. These nunataks represent
the summits of 2,000–3,000 m-high peaks
that are almost buried by ice.

Penny, and numerous smaller ice caps and valley glaciers (some tidewater) in the
eastern fjord country. Devon Island is dominated by a large ice cap at its eastern
end, which in places flows directly into the sea. Ellesmere Island has the most
extensive ice cover in the Canadian Arctic, with several large ice caps and icefields
discharging through many tidewater glaciers in the north and east. An ice shelf, the
Ward Hunt, is also developed on the north coast. Axel Heiberg Island has two main
ice caps, but here few glaciers reach the sea.

   In terms of area, the Inland Ice of Greenland is by far the most important in the
Arctic, representing over 80% of the total ice cover within the region. The estimated volume of 2.6 million km^3 represents 12% of the world's glacier ice which, if it
all melted, would result in a world-wide rise in sea level of 6 m, causing catastrophic
flooding of many of the world's major cities and low-lying countries. The thickness
of the ice sheet reaches 3,000 m in the south-central part. Numerous other local
glaciers, ice caps and valley glaciers comprise around 5% of the glacierised area, but
their volume is less than 0.5% of the total for Greenland.

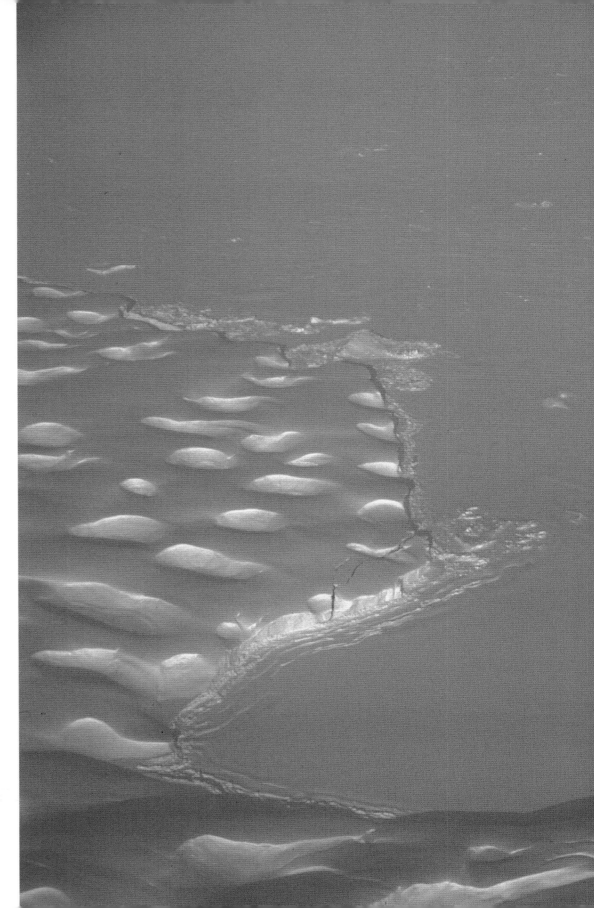

The southeastern sector of the Prince of Wales Ice Cap in Ellesmere Island, in the area of Cape Mouat, forms unbroken ice cliffs tens of kilometres in length where it reaches the waters of Baffin Bay. The frozen ocean is covered with smooth sea ice, punctuated with occasional icebergs, and the ice-cap surface contains a series of hollows typical of slow-moving or stagnant ice flow.

The main glacier-covered areas in the Eurasian Arctic are in Svalbard (60% ice-covered), Franz Josef Land (80% ice-covered), Novaya Zemlya (30% ice-covered) and Severnaya Zemlya (60% ice-covered). In each of these areas, extensive inland icefields and ice caps descend to sea level as valley glaciers form wide ice fronts from which many icebergs are produced. Further east, a few small glaciers exist in the De Long archipelago, but the New Siberian islands and Wrangel Island contain no glaciers.

Glaciers and climate: the health of glaciers

The state of health of a glacier, that is, whether it is growing or decaying in area and thickness, can be addressed through comparing the amount of snow received each winter with the amount of snow and ice that is lost by the end of the following

Tidewater glaciers drain from an icefield in East Greenland into the adjacent ice-infested waters. Crevasses parallel to the ice margin will eventually provide the lines of weakness along which icebergs are produced. Medial moraines, formed by surface debris, are responsible for the dark stripes on the ice surface. The photograph was taken from about 11,000 m, from an Airbus 340 aircraft on a commercial flight between London and Los Angeles.

A floating tongue of glacier ice protrudes from the eastern margin of the Academy of Sciences Ice Cap on Komsomolets Island in Russian Severnaya Zemlya. Large numbers of tabular icebergs are calved from the tongue, and are embedded in winter sea ice. The name Komsomolets translates to Young Communist League in English, and attests to the political zeal of Ushakov, who first explored the Severnaya Zemlya archipelago in the early 1930s.

summer through melting and subsequent runoff. Meltwater from snowmelt and slush may also contribute positively to the mass balance of a glacier. As this slush refreezes in winter it forms 'superimposed ice', adding to the mass of the glacier. Indeed, some glaciers in the Arctic islands rely on superimposed ice for continued nourishment. The sum of inputs and losses of mass in a given year is called the net mass balance of a glacier. If more snow falls in winter than melts during summer, the glacier has a positive mass balance. It will, therefore, grow. Conversely, if summer losses exceed winter accumulation, the glacier will become thinner and probably recede if a net loss of mass continues for a number of years. Many glaciers in the Arctic islands have been thinning and receding over the past hundred years or so, in response to the end of a cool period known as the 'Little Ice Age', that lasted several centuries.

Mass balance is usually measured by drilling a series of aluminium stakes into the glacier surface, and then measuring the height of the surface relative to each stake, first at the end of winter to obtain the amount of snowfall, and then again at

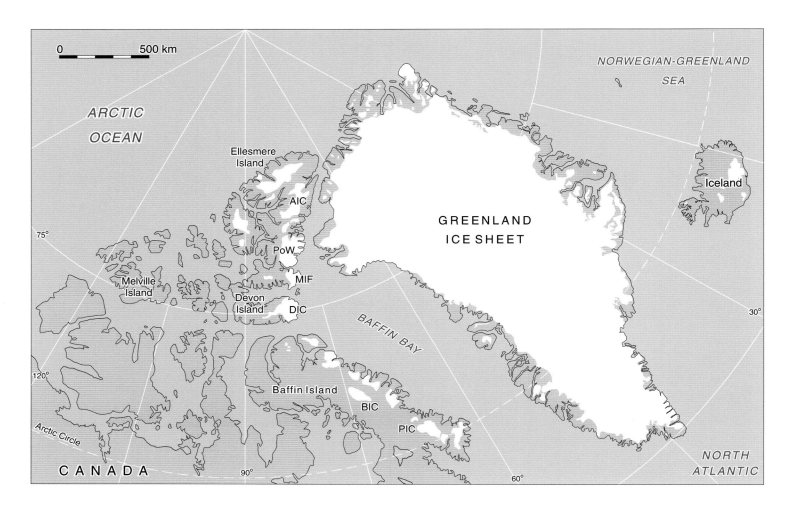

Map of the glaciers, ice caps and ice sheets in the eastern Canadian Arctic and Greenland. AIC is Agassiz Ice Cap, BIC is Barnes Ice Cap, DIC is Devon Ice Cap, MIF is Manson Icefield, PIC is Penny Ice Cap and PoW is Prince of Wales Ice Cap.

the end of summer to measure how much snow and underlying ice has been lost. For a glacier that ends in the sea, a further amount of ice is also lost by the production of icebergs. This quantity has seldom been measured, and requires knowledge of both ice thickness at the ocean margin and the speed of ice flow close to the terminus.

Glacier flow and surging

The closest analogy for the flow of ice is that of the slow deformation of a metal that is close to its melting point. Both metals and ice are solids, rather than liquids, and ice flow is through the deformation of crystals. Glacier motion resulting from

Table 4.1. *Areas (in km²) covered by ice on the Arctic islands*

Greenland	total ice-covered area	1,802,600
Ice sheet or 'Inland' ice		1,726,400
Independent glaciers and ice caps		76,200
Canadian Arctic islands	total ice-covered area	151,057
Ellesmere Island – glaciers and ice caps		80,000
– floating ice shelves		500
Baffin Island		37,000
Devon Island		16,200
Axel Heiberg Island		11,735
Bylot Island		5,000
Coburg Island		225
Melville Island		160
North Kent Island		152
Meighen Island		85
Svalbard	total ice-covered area	36,591
Spitsbergen		21,767
Nordaustlandet		11,309
Edgeøya		2,130
Ice caps and glaciers on other islands		1,385
Jan Mayen	total ice-covered area	115
Russian Arctic islands	total ice-covered area	55,635
Novaya Zemlya		23,600
Severnaya Zemlya		18,300
Franz Josef Land		13,735

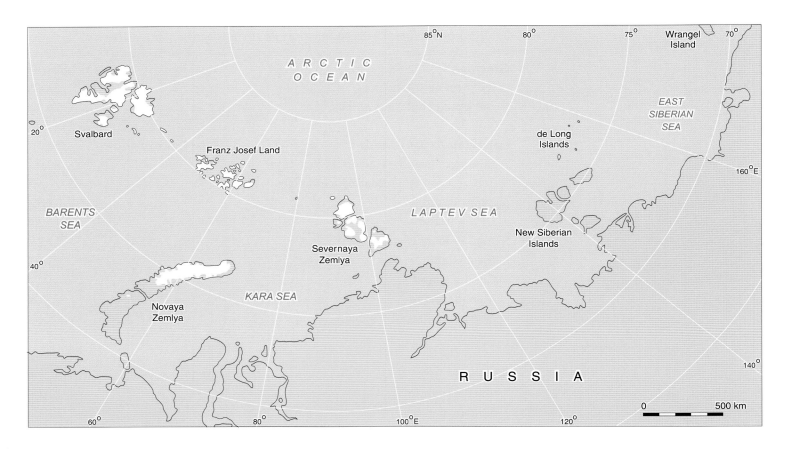

Map of the glaciers and ice caps in the Eurasian Arctic archipelagos.

this slow internal deformation is typically a few metres per year. The ice flows downslope under the influence of gravity. However, ice flow is controlled mainly by the slope of the ice surface rather than the bed, and for this reason ice near the base of a glacier can flow uphill, over obstructing bedrock ridges.

Observations of the speed of glaciers, made both by field surveys and from satellite measurements, show that ice flows at rates ranging from less than a metre to several kilometres per year. A number of large fast-flowing outlet glaciers drain the interior of the Greenland Ice Sheet, and Jakobshavns Isbrae in West Greenland has a flow rate of 7 km per year, the fastest observed anywhere in the world. Speeds such as this are not produced by crystal deformation alone. It is the presence of water, and sometimes water-saturated sediments, at the glacier bed that provides the necessary lubrication for fast flow.

But, how can the temperature beneath hundreds, and sometimes thousands of

Towards the end of summer, a small snow patch is all that remains of the winter snowfall on the low-lying Wrangel Island, in the Eurasian Far East of the Arctic (Photo: M. Saarnisto).

metres of ice reach the melting point to allow water flow? The coldest place on a glacier or ice sheet is close to the surface, because the ice is in contact with the cold atmosphere. With increasing depth through the ice, the temperature actually increases. This is because heat from the Earth's interior, called geothermal heat, and heat generated by ice-crystal deformation, act to warm the ice. Friction, caused by the glacier sliding over the bed, also plays a role in warming the ice at depth. The further from the cold surface, the higher the temperature, a phenomenon observed in drill holes several kilometres deep through the Greenland Ice Sheet. Where the melting point is reached at the glacier bed, water may be located in several places. It can be in channels cut into the bedrock or the overlying ice, in a thin film, or held within soft sediments immediately beneath the ice. In each case, friction at the glacier bed is reduced and ice can then move at speeds much faster than through ice deformation alone.

Many glaciers in the Arctic islands have receded and thinned over the last hundred years or so, in response to climate warming. The parallel sediment ridges or moraines at the side of this East Greenland glacier in Kangerlussuaq Fjord mark its former margins and demonstrate that it has thinned recently.

A scientist using survey methods to measure the speed of the Austfonna ice cap on Nordaustlandet in eastern Svalbard. A snowmobile and sledge are used to move about on the ice cap. The aluminium poles, topped with flags in order to be spotted from a distance of several kilometres, are used to observe both ice motion and snow build-up.

A small outlet glacier flows north from the Grinnell Ice Cap in southernmost Baffin Island into Frobisher Bay. This ice cap, together with the neighbouring Terra Nivea, are the most southerly ice caps in eastern Canada. The undulating and crevassed surface of the glacier suggests that it is flowing over a series of bedrock ridges.

In addition to being important components of the landscape and modifiers of the local climate, glaciers receive attention because of their sometimes unpredictable behaviour. Many Arctic glaciers undergo catastrophic advances known as surges, sometimes of several kilometres in a few months, which are unrelated to any climatic changes. Surge-type glaciers remain inactive for many decades, punctuated by much shorter periods of fast flow. This shift from stagnation to rapid flow and terminus advance is thought to be associated with the buildup of water to lubricate the glacier bed or to a change in the state of subglacial sediment from frozen to wet. Surging glaciers may dam major rivers or even fjords temporarily. Many other glaciers trap water in the form of internal reservoirs or ice-dammed lakes, the sudden release of which causes catastrophic flooding.

The advancing 30 m-high cliff of Thompson Glacier on Axel Heiberg Island, illustrating the large amount of debris incorporated into the ice from the bed, and a powerful meltstream from the adjacent White Glacier.

Surface features on glaciers

The surface of a glacier bears witness to the forces generated during ice flow, the way in which debris is transported, and the role played by meltwater. When ice is stretched, as it goes round a bend, over a rock bump or step in the bed, or on entering the sea, it fractures in brittle fashion, producing crevasses. Crevasses are a major hazard to glacier travel, especially in the accumulation area, where they may lurk largely hidden beneath snow. Crevasses are commonly several metres wide and a few tens of metres deep, although few direct measurements have been made. They usually form distinct geometrical sets and, where several intersect, travel is usually impossible. The most intense crevassing occurs when glaciers surge. At such times,

Many glaciers on the Arctic islands undergo short-lived rapid advances known as surges. Here are two views of Fritjhovbreen in central Spitsbergen during its 1998 surge. The upper photograph is an aerial view of the heavily crevassed surface; the lower one shows the terminal cliff about 20 m high as it advances across land at a rate of several metres per day.

A series of crevasses, almost parallel in their orientation, have formed on the surface of an Ellesmere Island glacier in response to the stresses induced by ice flow over a rough bed.

the whole glacier surface may break up. Most accidents on Arctic glaciers are a result of falls into crevasses, and rescue from a crevasse demands sophisticated rope-work.

In addition to brittle structures, there are structures arising from ice deformation during compression and shearing, including foliation and folding. Foliation is a layered structure made up of ice with different crystal types. It is a useful structure for navigation in mist, as it is commonly orientated parallel to the valley sides, and thus points the way up or down the glacier.

Debris is often exposed at the glacier surface as linear streaks parallel to the flow, called medial moraines. Some medial moraines are initiated by the convergence of two streams of ice, whilst others melt out from the ice near the glacier snout. The way in which debris is incorporated in the ice, and subsequently modified and deposited, is a very complex process related to folding, compression and shearing.

Summertime on Arctic glaciers rings to the tune of running water. As the snow-

Where different glaciers join together, the debris at the margins merges to form medial moraines, represented by stripes of angular boulders on the glacier surface, as here on Edward Bailey Gletscher in Renland, East Greenland.

pack melts, slush swamps form before the water is released in sheet-floods or in channels. Meltwater channels, cut into the ice surface, commonly meander or follow the various structures. Each year the same channels may be reoccupied, so they can attain depths of several metres. On many glaciers, these streams plunge down holes known as moulins that may be tens of metres deep. Their smooth, slippery walls make meltwater channels a further hazard on the ice surface. The thermal structure of Arctic glaciers is such that water often fails to reach the bed, but is forced to the sides. Thus, many Arctic glaciers have lateral streams along their flanks.

The glacial landscape

Most Arctic glaciers are undergoing steady recession in today's climate, at rates ranging from a few metres to hundreds of metres per year. As they retreat, they leave behind a record of erosion and deposition that scientists can interpret in the

Two geologists survey the lateral ice cliffs and marginal meltwater stream of Wordie Gletscher in East Greenland. The sub-zero temperatures of the ice force englacial meltwater to the margins before it can reach the bed. Here the stream undercuts the cliff, causing ice falls which temporarily block the stream.

context of climatic change spanning thousands of years. Moreover, since more temperate regions, such as much of North America and northern Europe, were affected by glaciation during the last and preceding ice ages, we can learn much about environmental conditions in these areas by examining the processes operating near glacier margins in the Arctic today. We can also apply our knowledge of depositional processes and products in Arctic regions to evaluate sand and gravel resources, groundwater flow and waste disposal sites in formerly glaciated terrain.

Glaciers and ice sheets affect the landscape through both the initial erosion and transport of debris quarried from the glacier bed, and the subsequent deposition of this eroded debris. In general, mountainous areas tend to have a landscape characterised by glacial erosion, whereas lowland regions are often dominated by the deposition of glacier-derived sediments.

Glacial erosional landforms
Glacier recession since the Last Glacial Maximum about 18,000 years ago has, with few interruptions, progressively exposed more and more land in the Arctic,

revealing a wide range of both dramatic and subtle erosional landforms. The most spectacular features of glacial erosion are glacial troughs, a term embracing both valleys and fjords (see Chapter 7 for the latter). In cross-profile, glacial valleys are steep-sided and flat-bottomed, commonly but often misleadingly characterised as U-shaped. Rivers, by contrast, tend to produce valleys that are 'V'-shaped. This is because glaciers erode bedrock throughout a valley, as they are in contact with both its floor and side walls, whereas rivers flow in a channel on the valley floor and do not erode the valley sides directly. Glacial valleys have long straight reaches, and over-deepened basins, either occupied by lakes or filled by sediment. Braided rivers and mosquito-breeding bogs commonly occur on the valley floors.

Small glaciers, perched on steep mountainsides, tend to erode out armchair-shaped hollows up to a few kilometres across. These features are termed cirques

A series of cirque glaciers fill rock basins cut into the mountains of the East Greenland coast on the north side of Kangerlussuaq Fjord.

The crevasse-free surface of a small receding valley glacier, Austre Lovenbreen, in northwest Spitsbergen. The main obstacles to progress in such terrain are deeply incised meltwater streams which form the distant lines across the glacier surface. In the background is a knife-edge ridge, culminating in a pyramidal peak or 'horn'.

(from the French) or, more locally, corries in Scotland or cwms in Wales. These characteristic features of a glaciated mountain landscape often remain ice-filled in the islands of the Arctic. Although cirques are usually associated with relatively high altitudes, many in the Arctic islands are close to sea level.

Headward erosion of two back-to-back cirques ultimately results in the intersection of the backwalls to produce narrow pinnacled ridges called arêtes (from the French). If three or more cirques erode backwards, the arêtes rise to a sharp pointed peak referred to as a horn (from the German). Such features are actively forming today in areas like Spitsbergen, Greenland and Ellesmere Island, and demonstrate how similar features formed in areas no longer occupied by ice.

More subtle forms of glacial erosion may also be found, especially on bare bedrock surfaces. Of these, scratch marks or striations and polishing are the most important. They testify to the abrasive characteristics of glacier ice laden with rock debris as it slides over the bed. Abrasion is supplemented by bedrock fracturing (plucking), especially on the downglacier flanks of rock obstacles. Whaleback-

As a glacier slides over its bed, it picks up rock fragments which act as abrasive and scouring agents. As a result bedrock surfaces, such as this marble outcrop on Ossian Sarsfjellet in northwest Spitsbergen, commonly bear polished and scratched ('striated') surfaces, as well as various gouge-marks.

The same process also gives rise to striated stones, such as this limestone cobble near Kongsvegen in northwest Spitsbergen. Not all rock types are susceptible to scratching, however, but limestone fragments are commonly striated because of their softness.

Angular debris, derived from rockfall from the steep mountainside behind the photographer, rests on the surface of a small cirque glacier on the Meta Incognita Peninsula in southermost Baffin Island, Canada. Looking down the glacier, the fjord of Frobisher Bay is seen. It is free of sea ice in this late summer picture.

shaped forms, with smoothed and striated upglacier-facing surfaces and plucked downglacier faces, are known as roches moutonnées; this term has nothing to do with sheep, but with the wigs that were fashionable in eighteenth century France.

Glacial sediments and depositional landforms

Glaciers transport debris both at the bed and at the surface. Material at the bed is derived from the erosive action of ice itself, whereas debris at the surface is more often derived from rockfalls and dirty-snow avalanches from valley walls or from nunataks exposed above the ice surface. Stones from the glacier bed usually have somewhat smoothed edges, and may also be striated. By contrast, boulders derived

Many different processes operate at modern glacier margins, not only direct deposition from the ice. Because of the availability of meltwater in summer from exposed and buried ice, sediment is especially prone to flowage. Here we see glacial deposits resting on stagnant glacier ice several kilometres from the active margin of Kongsvegen in northwest Spitsbergen. As the ice melts, the debris cover collapses and flows downslope as a slurry.

A wide range of depositional landforms is associated with glacier margins. Here are a series of stacked slabs of debris, making up a group of 'hummocky moraines'. In this example, adjacent to Kronebreen in northwest Spitsbergen, the moraines are the result of deformation within the glacier. Similar landforms are widely developed in areas no longer occupied by ice such as Scandinavia, the British highlands and North America.

from rockfall are typically angular in shape, illustrating their origin through frost action.

Debris release is often accompanied by stagnation of the edges of the glacier. The melting ice provides a ready supply of water to saturate the sediments, resulting in considerable debris-flowage. As ice melts, debris is released, especially at the margins of a receding glacier, and a characteristic deposit called till is produced. In addition, the glacier produces a distinctive collection of depositional landforms. Glaciers carry debris of all sizes, from the finest clays to boulders as large as houses. A walk across the sediments just beyond the terminus of a glacier illustrates this very clearly, as ridges and mounds of material, ranging in 'particle' size from large boulders to clay, will be followed by hollows containing clinging mud and occasionally quicksand. To add to the obstacles, large meltwater streams will usually need to be crossed.

The largest depositional landforms are moraines – ridges of sediment released by the ice along the valley sides (lateral moraines) or pushed up by the ice in front

Glacial erosional landforms are major features of the British mountain landscape. Classic landforms include 'cirques', also known as corries (in Scotland) and cwms (in Wales). They often have glacier-carved rock basins filled with small lakes called tarns. This is Cwm Cau on Cadair Idris, one of the best examples in Wales.

(end-moraines). Hummocky moraine is the product of deformational processes in the ice. The sediments involved in the construction of moraines are varied – they can include material of marine, river or direct glacial origin. The size of the ridges, a few tens of metres high, is deceptive as far as the end-product is concerned, as the moraines often have an ice core, which melts only very slowly with time, and can be hundreds or even thousands of years old.

Other depositional landforms are the product of fluvial (river-flowing) processes in contact with the ice. At the glacier margins, lateral meltwater streams deposit sand and gravel which, when the ice recedes, are left as benches called kame terraces, perched above the valley floor. Meltwater tunnels beneath the ice can be clogged with similar sediment which, when the ice melts, reveal upstanding ridges called eskers.

The results of glacier erosion are seen clearly in Yosemite National Park in the Sierra Nevada mountains of California, USA. Ice, last present about 18,000 years ago, has smoothed the granite rock and left behind large numbers of isolated boulders when it decayed.

Comparisons with other areas

Both the erosional and depositional landforms produced by glaciers are distinctive. This allows them to be recognised quite easily – not only in the Arctic islands, where such processes are active today, but also at lower latitudes where ice has been important in sculpting the landscape during colder periods in the past. The upland areas of Britain and the Rocky Mountains of North America, for example, both provide classical illustrations of past glacial activity. Scientists, therefore, continue to study the processes and patterns of glacial erosion and sedimentation at a number of locations within the Arctic islands, and to apply this knowledge to the interpretation of landscapes elsewhere which bear the imprint of past glacial activity.

Chapter 5 Icebergs and sea ice

What are icebergs and sea ice?

The waters around the island archipelagos of the Arctic are infested by ice in two forms – icebergs and sea ice. Both these types of floating ice can take on a variety of shapes and are among the many spectacular natural features to be seen in Northern waters. Icebergs are derived from glaciers and larger ice caps that terminate in the sea as tall ice cliffs. Pieces of ice, which can vary in length from metres to kilometres, break off or calve from these cliffs to form the bergs. The ice in icebergs comes originally from snowfall on the glacier surface and is transformed into glacier ice as a result of pressure during burial. This ice may be hundreds or even thousands of years old by the time it calves into the sea. By contrast, sea ice comes from sea water and is formed when the sea surface freezes. Individual pieces of sea ice are called floes, and large areas of sea ice are referred to as pack ice. Icebergs and sea ice are less dense than water, because of the presence of air bubbles that are trapped within them during formation. This density difference gives them some buoyancy, though about 90% of their bulk is hidden below the water surface. Both types of ice present hazards to navigation in the polar seas and, in the past, provided significant natural barriers to the exploration of the Arctic islands. However, winter sea ice provides a ready means of surface transport on foot, on ski, by dog-team or by snow-scooter although, as many failed attempts to reach the North Pole testify, numerous obstacles may need to be overcome.

The seaward margin of an ice-cap outlet glacier in eastern Ellesmere Island produces icebergs, which are trapped in the smooth winter sea ice covering the adjacent fjord.

Large stranded floes of multi-year sea ice derived from the polar pack are sometimes caught up in the current that sweeps around the southern tip of Spitsbergen. Occasionally, these floes drift northwards a considerable distance and fill up the west coast fjords, resulting in difficult or impossible navigation for small vessels even in midsummer, as here in Bellsund.

Iceberg dimensions and drift

Icebergs are produced by any glacier that reaches the sea. The thickness of an iceberg depends, initially at least, on the thickness of the glacier from which it is derived. However, some ice margins produce large, flat-topped or tabular icebergs with deep below-water parts known as keels. These tabular icebergs are often derived from ice margins which are themselves floating. At other ice margins the terminus is aground, rather than floating, and may also be broken by crevasses. These margins tend to produce many smaller icebergs of more irregular shape.

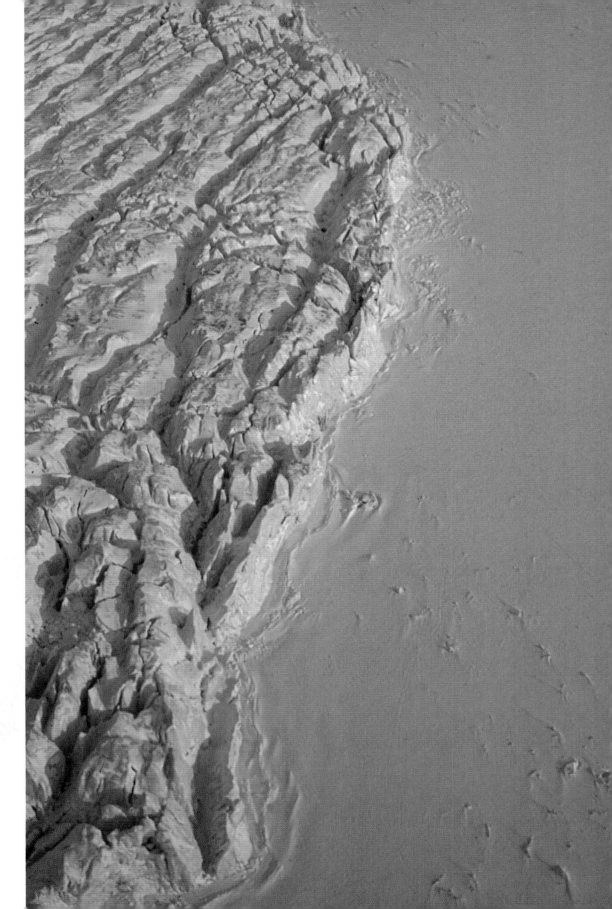

The crevassed terminus of a Spitsbergen glacier is viewed from a helicopter hovering above. Periodically, the ice collapses along the line of a crevasse and falls into the adjacent fjord to form small icebergs, often of irregular shape. In this spring photograph, the fjord waters are covered with sea ice, which is smooth except close to the glacier margin where blocks of ice have fallen.

The potentially dangerous underwater keel of an iceberg protrudes beyond its exposed upper surface. The above-water ice cliff is about 15 m high. Ships usually give icebergs a wide berth, because of their hazardous nature and the possibility that they may overturn without warning.

The largest icebergs in the seas around the Arctic islands are a few kilometres in length and up to about 600 m thick. However, these are small, in length although not in thickness, compared with those that break off the ice shelves of Antarctica. The largest Arctic icebergs are produced mainly by outlet glaciers from the Greenland Ice Sheet, which often flow at speeds of several kilometres per year. Some of the largest Arctic icebergs are produced from Jakobshavns Isbrae in West Greenland and from Daugaard–Jensen Gletscher on the east coast. In the Eurasian Arctic, the glaciers of Franz Josef Land and Severnaya Zemlya are the main sources of large tabular icebergs. However, all the observer from the deck of a passing ship would see is the uppermost few tens of metres of these large bergs, their underwater keels glimpsed only as blue forms beneath the sea surface.

Once icebergs have broken off from their parent glacier, they are free to drift under the influence of ocean currents and the wind. Currents are a stronger influence on the path taken by icebergs than wind direction, because of the huge underwater bulk of the bergs. Large icebergs can travel very long distances away from the glaciers that produced them. Icebergs carry debris that is derived from

A fast-flowing outlet glacier on Ellesmere Island is producing large icebergs, which break off along crevasses formed parallel to the ice front. The very flat nature of the ice margin indicates that the glacier tongue is probably floating on the fjord waters. As well as large tabular icebergs, very many small bergy bits are also calved from this active glacier.

erosion by their parent glacier, and rocks bearing the distinctive geology of their place of origin have been found by scientists hundreds of kilometres away from their source. Thus, coal fragments eroded by glaciers from Spitsbergen and even Franz Josef Land have been found on the floor of the Norwegian Sea about 1,000 km to the south of these archipelagos.

Along with the larger bergs comes large quantities of 'brash ice' – masses of small fragments of glacier ice a few metres in length or less. This type of ice tends to be associated with glaciers grounded on the sea floor, and is the result of faster forward motion towards the top of the glacier cliff, so that ultimately the face leans

Left: Icebergs in Antarctic Sund, East Greenland, derived from outlet glaciers from the ice sheet over 100 km away. The beached iceberg is typical of the smoothed, beautiful forms like modern sculptures, that are stranded at low tide.

Near-vertical aerial view of a heavily-crevassed iceberg in Nordvestfjord at the head of the Scoresby Sund fjord system in East Greenland. The iceberg is several hundred metres across, and its rough surface indicates that the iceberg has been calved from a fast-flowing outlet glacier of the Greenland Ice Sheet. Bergs such as these may have underwater keels over 500 m thick.

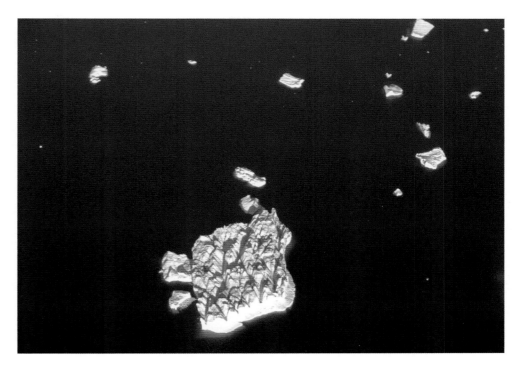

The combination of fog and icebergs is a major hazard to coastal navigation, as here in Kongsfjorden, northwest Spitsbergen. The black guillemot provides a scale for this berg, viewed from a passing rubber

Floating icebergs often disguise the fact that they are major conveyors of debris into the marine environment. This unusual aerial photograph shows an overturned berg, displaying a remnant of the debris-rich sole of Vestfjord Gletscher in East Greenland.

forward and simply collapses. Smaller icebergs like this are called 'bergy bits' and often wallow low in the water. Bergs of relatively high density that are almost submerged are known as 'growlers'.

As icebergs drift they undergo melting, at a rate which is dependent on the temperature of the sea. Icebergs that have been in the water for a long time can take on a wide variety of shapes as a result of the melting of both their above-water and below-water parts. Indeed, as they melt, their centre of gravity may shift, and areas that were underwater may be exposed to view. Former waterlines, arches and pinnacles can sometimes be seen. In fact, the variety of shapes that icebergs can take is almost infinite. Melting can also make bergs unstable and overturn, and fragmentation can take place when storm waves, or even the wake of a passing vessel, disturb them. Many icebergs fail to reach the open sea, but become stranded on shallow banks, where they break up and melt. Bergy bits may become stranded on beaches at high tide, providing attractive foregrounds to fjord scenes. As the bergs move up and down the beach with the tides, they leave behind distinctive imprints or push ridges.

Why are icebergs sometimes blue?

Icebergs are a very impressive sight in the Arctic seas, not only for their myriad of shapes but also for their colour, which is sometimes a clear blue. This blue colour is an optical effect, produced by the absorption of red light and the continued transmission of light in the blue part of the spectrum. It takes an appreciable thickness

As icebergs melt, they acquire castellated forms and caves may develop, as in this solitary berg in Antarctic Sund, East Greenland. Note how slight tilting of the berg has occurred, exposing the former waterline on the left.

An iceberg in Scoresby Sund, East Greenland, shows a series of parallel waterlines, demonstrating that it is unstable and shifting its centre of gravity as it melts and fragments.

of pure ice in order to absorb enough red light to make an iceberg look blue, and for the light to travel far through an iceberg there must be few air bubbles or debris particles. However, ice often contains air bubbles, trapped when the original air pathways in the snowfall making up glaciers were closed off to form ice. Air bubbles in the ice scatter incoming light and reflect much of it back after only a short travel path within the ice. As a result, little absorption of red light occurs, and the ice appears white rather than blue. Blue icebergs are, therefore, those where the ice is particularly pure and free of internal scatterers such as bubbles or debris.

As the tide goes out, bergy bits become strewn along this beach in Signedalen, an inlet of Krossfjorden in northwest Spitbergen, forming a dramatic foreground to a beautiful fjord scene. The constant crackle from bergs like this is due to the release of entrapped air as the ice melts.

Iceberg hazards

Iceberg collisions with shipping are popularly viewed as a major hazard in the Arctic seas. The iceberg that sank the *Titanic* in 1912 with the loss of 1,503 lives was almost certainly derived from the Greenland Ice Sheet, and had drifted several hundred kilometres before the collision. Today, large icebergs can be identified from many kilometres distant by the sophisticated radar systems used by most vessels operating in polar waters. More problematic are the last remnants of these bergs, the so-called 'growlers', a few metres across, that can be almost awash and hidden from view in a stormy sea. It is for hazards like this that ships designed to

As bergy bits are jostled by the tide near beaches, they create juddermarks on the soft sediment. These marks, resembling the trace of a tracked vehicle, are shown with the culprits, now stranded at high-water mark, on this beach near Austre Lovénbreen in northwest Spitsbergen.

operate in the Arctic seas are usually built with double hulls, so that they have two watertight barriers for protection in the event of damage by ice.

Icebergs also present a very significant hazard to structures on the sea floor. Their below-water keels, reaching up to several hundred metres below the sea surface, can gouge huge furrows in the sea bed when they drift into relatively shallow water. Side-scan sonars, scientific instruments that collect images of the sea floor similar to aerial photographs of the land surface of the Earth, show that large areas of Arctic continental shelves are disturbed or 'ploughed' by iceberg keels.

A team of divers from the Danish navy enters the icy waters of Kangerlussuaq Fjord in East Greenland to examine the hull of a ship that has been damaged below the waterline by collision with an iceberg. The divers wear dry suits, but still have their faces partly exposed to the almost freezing water.

Thus, any engineering structures, such as the well-heads and pipelines associated with oil and gas fields, are susceptible to damage by icebergs. Sea-ice floes, too, can have submarine keels of up to 10 or 15 m, and ploughing can also be a result of sea-ice action in very shallow water. The problem of ice ploughing is already recognised as a considerable hazard on, for example, the Grand Banks off Newfoundland in eastern Canada. Sea-ice keels could also represent a considerable problem in the exploitation of the extensive gas reserves in the southeastern Barents Sea close to the southern part of the Novaya Zemlya archipelago.

The track of an icebreaker through newly-formed sea ice can be seen clearly in this picture taken in late September, as the Arctic autumn closes in on Scoresby Sund, East Greenland. Taking off from the heli-deck of the icebreaker, scientists prepare to measure the size of icebergs in Nordvestfjord in the Scoresby Sund fjord system. In the distance is one of the huge icebergs derived from the large and fast-flowing glaciers that are fed from the Greenland Ice Sheet.

Left: The 118 m-long German icebreaker *Polarstern*, silhouetted against the sun in Scoresby Sund, East Greenland. Her radar and communications antennae can be seen, along with many large icebergs drifting down the fjord towards the Greenland Sea.

Sea-ice growth and decay

Sea ice forms as the surface of the sea freezes during late autumn and winter in the Arctic. It is interesting to note that the freezing point of sea water is significantly lower than that of fresh water, and sea water will not freeze until a temperature of $-1.8\ ^\circ$C is reached. This is because the high salt content of sea water depresses the freezing point, and is the reason why we salt roads in an attempt to prevent ice formation. The salt within sea ice also makes the ice appear sticky when skiing. Relatively calm water is usually required for freezing, because waves will tend to mix the very cold surface waters with slightly warmer water from below.

Sea ice usually grows to between one and a few metres in thickness, and individual floes may be anything from centimetres to kilometres in length. Almost all fjords and inlets in the Arctic islands have a continuous cover of sea ice, known as 'shorefast ice' or simply 'fast ice', during winter and spring. The shorefast ice moves up and down with the tides, which are usually 1 or 2 m in height, but reach up to 10 m in the fjord of Frobisher Bay in Baffin Island. A tide-crack forms between the

A smooth blanket of shorefast sea ice, about 2 m thick, covers the surface of Dobbin Bay in eastern Ellesmere Island during the Arctic spring. A series of icebergs, calved from the nearby Eugenie Glacier, are trapped within the sea ice at this time of year. The icebergs and sea ice will drift into the waters of the Kane Basin after breakup of the shorefast ice in summer.

continuous plate of shorefast ice and the coast as the ice flexes with the rise and fall of the tide, and can form an obstacle for travellers. The breakup of sea ice often takes place only in June or early July, and dispersed drifting floes remain in many fjords throughout the summer season. In some years, especially those with few storms and particularly low summer temperatures, the ice remains through the summer. This is the case in some sheltered fjords in East and North Greenland, where such multi-year shorefast ice, known as *sikussak* in Greenlandic, traps large numbers of icebergs within it.

In the open seas beyond the protection of the Arctic islands, sea ice is more broken, and is composed of floes of all sizes. Most are less than about 3 m in thickness. However, where wind and wave action force floes together under pressure, they may ride up over one another to form ridges. Thus, the surface of the Arctic

The outer coastal fringe of East Greenland retains ice floes even in late summer. This scene, showing the mountains reflected in the calm waters between floes, is from an area known as Canning Land.

The shorefast sea ice formed in winter in the inner part of Hornsund, southern Spitsbergen, breaks up in early summer as a series of rectangular ice floes. The tidewater glacier in the background is also calving icebergs into the fjord.

The British research vessel *James Clark Ross* navigates through sea ice in the Greenland Sea about 150 km offshore from the East Greenland coast.

pack ice is not necessarily smooth. This is one of the reasons why explorers trying to reach the North Pole make such slow progress, as ridges can reach several metres in height and have a broken and irregular shape. Although pack ice in the Arctic Ocean is constantly on the move, much of it is trapped, and individual floes can grow year by year. It is estimated that some sea ice can reach ages approaching ten years before disintegrating. Old Arctic Ocean pack ice that escapes into the adjacent seas is often recognisable on account of its size. In summer it may, for example, drift unexpectedly into the normally ice-free fjords of western Spitsbergen, by floating from the northern Barents Sea round the island's southern cape (Sørkapp), where it gets caught up in the continuation of the North Atlantic Drift. Events like this can be a severe hindrance to coastal navigation. As with icebergs, sea ice also can become stranded in shallow water and on beaches, where it may attain very unusual forms.

Another problem for travel on sea ice is that the Arctic pack is constantly drifting under the influence of winds and currents, with wind often being the dominant force. The pack can be moving in one direction, whilst ships trapped in the ice or people walking on the surface can be moving in another. Fridtjof Nansen, the Norwegian explorer, used the drift of sea ice to his advantage, bringing his ship *Fram* closer to his target, the North Pole. The members of Andrée's balloon expedition to the North Pole in 1897 also found this out to their cost when attempting to walk back to the safety of Spitsbergen after their balloon crashed on the Arctic pack ice. They did not succeed in racing against the drift of the sea ice, and their bodies were found in 1930 on the remote White Island, well to the east of Spitsbergen.

Sea ice, climate and ocean circulation

Sea ice covers an area of almost 16 million km^2 of the Arctic seas during its maximum extent in winter. Each summer the pack ice melts and shrinks in area, and, at a minimum, extends over about 9 million km^2. The sea ice is an important

As the large ice floes from the polar pack disintegrate and become stranded at low tide, they take on a variety of elegant forms, as here in Van Keulenfjorden, western Spitsbergen.

Map of the distribution of sea ice in
the Arctic. The maximum winter
extent and the minimum summer
extent of sea ice are shown.

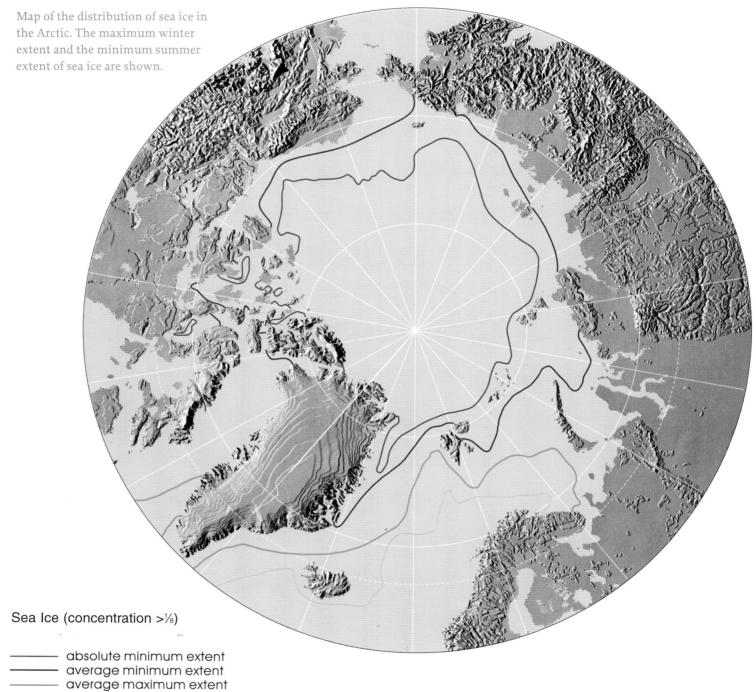

Sea Ice (concentration >⅛)

——— absolute minimum extent
——— average minimum extent
——— average maximum extent
——— absolute maximum extent

part of the climate of the Arctic, and indeed of the whole world, because it has a large effect on the way the surface of the Arctic is heated. Sea ice is highly reflective – we see this by the need to wear sunglasses when walking over snow or ice. Water, on the other hand, appears rather dark by comparison. Where sea ice covers the Arctic seas, much radiation from the sun is reflected back into the atmosphere. However, where open water is present, most of that radiation is absorbed and acts to heat the water. As the water heats up, more ice can be melted, and so on. This is called a positive feedback effect, and is one of the key reasons why scientists are monitoring both the extent and the thickness of Arctic sea ice today. If sea ice covers a progressively smaller part of the Arctic seas in a World that may be affected by greenhouse gases, then this positive feedback could be critical to the rate at which our planet warms up.

Icebergs and sea-ice floes of various sizes are photographed from the air in an East Greenland fjord. The thinnest sea ice, only some few centimetres in thickness, is relatively dark, and thicker sea ice and the larger icebergs appear brighter. The open-water areas of the photograph are darkest and, therefore, absorb more energy from the Sun.

Floes formed in the Arctic Ocean are much larger than those formed in fjords, and are thus readily distinguishable. Here, a few large floes, their underwater keels just showing through the clear water, are floating through the straight known as Hinlopenstretet, with Nordaustlandet, northeast Svalbard, in the background.

Another important aspect of the formation of sea ice is that, when water freezes, a large proportion of its salt content is rejected. Indeed, it is possible to melt old sea ice and drink the water. However, the sea water produced in areas where sea ice forms is both very cold and also of high salinity. The presence of so much salt produces some of the densest water found in the world's oceans, and this water sinks to the sea floor as a result. In fact, the very dense waters produced by sea-ice freezing in the Arctic are a vital factor in driving the circulation of the oceans, because these waters flow to the deepest parts of the ocean floor. Some areas of the Arctic seas are places of intense sea-ice production where upwelling or strong winds consistently keep the sea open year-round. Examples include the North Water at the head of Baffin Bay, and Storfjorden between Spitsbergen and Edgeøya in the Svalbard archipelago. New sea ice can, therefore, continue to form throughout the winter in such open-water areas, which are known as *polynyas* from the Russian language.

Sea ice and access in Arctic waters

Sea ice also presents a number of hazards to human activity in the waters around the Arctic islands. In the heroic era of nineteenth and early twentieth century exploration of the Arctic archipelagos, the presence of sea ice in many sounds and passages between islands made travel by sea difficult and dangerous. The British Navy, for example, sent several tens of ships to the Canadian Arctic during the first half of the nineteenth century in an attempt to find a navigable Northwest Passage as a route for trade with the Indies and the Far East. However, this period of exploration coincided with some of the coldest weather of the last few thousand years, which was known as the Little Ice Age. The ice conditions encountered by many of these ships were appalling. A number of vessels became trapped in sea ice as the short summers came to an end, and several were crushed when pressure built up in the sea ice as it was driven by strong winds. The great dangers inherent in these early explorations of the islands and passages in, for example, the Canadian north

A large Russian icebreaker, with hammer and sickle emblem visible under the bows, moves through shorefast sea ice in a sound within the Franz Josef Land archipelago. A nuclear-powered icebreaker has already cut an initial passsage through the ice, and the diesel-powered ship is following in line astern. These powerful icebreakers are the main means of resupplying food and fuel to the scientific and military bases in the Russian Arctic islands.

are exemplified by the remarkable story of Sir John Franklin, and his crews of the ships *Erebus* and *Terror*, who disappeared in the late 1840s and whose fate has never been finally established. By contrast, the Norwegian explorer Roald Amundsen was the first to navigate a ship through the Northwest Passage, and he did so only in 1903–06, when the climate had warmed up after the Little Ice Age and sea-ice conditions had become less severe.

More recently, the waters north of Siberia, passing immediately south of the Russian Arctic islands of Novaya Zemlya, Severnaya Zemlya and the New Siberian Islands, have been seen as a possible new route between Europe and the Far Eastern economies of Japan and Korea. This passage, known as the Northeast Passage in historical times and the Northern Sea Route today, is traversed in summer by Russian coastal vessels, and the entrances to the great northward-flowing Russian rivers, such as the Ob and Yenisey, are kept open in the warmer months by Russian

Sea-ice floes stretch away into the distance in this spring photograph of sea ice in the Kara Sea. Open-water areas are important to heat and moisture exchange between the atmosphere and ocean, and this is why clouds tend to form above such open-water leads.

nuclear-powered icebreakers, which can cut through several metres of ice. In fact, icebreakers have rather blunt bows, and break ice by riding up onto it and cracking it with their own weight. Neither the Northern Sea Route nor the Northwest Passage are seen as commercially viable shipping lanes today, but if sea ice were to thin and recede over the next century as a result of global warming, then this view could be subject to reassessment.

We can see, therefore, that sea ice in polar waters is generally regarded as a nuisance or a hazard to human activity. However, the frozen surfaces of many Arctic fjords also provide straightforward routes for travel by dog-sledge, ski or snow-scooter during the winter and spring. This is because freezing in such sheltered inlets produces shorefast sea ice, which forms a solid, unbroken surface up to several metres in thickness that does not usually break up until about June. Indigenous Inuit hunters have traditionally used shorefast ice both to hunt from and as a highway between settlements and camps. Now, tourist groups in the Arctic islands are increasingly using these easily traversed routes to gain access to areas that can be reached in high summer only by boat or helicopter. Arctic wildlife, such as polar bears, seals and walrus also use the sea ice for hunting or simply for basking. Open-water areas within the winter sea-ice cover, known as leads, are areas where wildlife congregates prior to ice breakup.

Chapter 6 Frost action

In the year 1577, the explorer Sir Martin Frobisher embarked on his second voyage in search of the Northwest Passage through the Canadian Arctic archipelago. During his voyage he observed that the ground was frozen throughout the year – to depths of 'four or five fathoms, even in summer'. The frozen condition 'so combineth the stones together that scarcely instruments with great force can unknit them'. Despite these early observations, it was only in the late nineteenth century that frozen ground phenomena became the subject of scientific investigation. The triggers for these studies were the engineering problems faced in the construction of the Trans-Siberian Railway across extensive areas of frozen tundra, and the difficulties in establishing the infrastructure for gold exploitation in Alaska and the Yukon.

Although much of the Arctic islands is covered by glacier ice and thus strongly influenced by glacial processes, we now know that most of the remaining land surface is under the influence of frost action, producing a landscape referred to as the periglacial zone. Ground frozen permanently to a great depth is the main feature of the periglacial zone, and is known as permafrost. Periglacial processes are dominant in ice-free islands, such as the New Siberian Islands in Arctic Russia, and Banks and Victoria islands in the Canadian Arctic. The sub-zero temperatures for most of the year mean that the ground is frozen solid to depths of hundreds of metres, inhibiting the free flow of water. Only the surface layer thaws out and, in such waterlogged conditions, plants have had to become specially adapted, whilst the construction of buildings poses particular problems. In the surface layer,

Talus or rockfall debris has formed at the foot of steep cliffs in Dobbin Bay, Ellesmere Island, Canada. The talus cones are snow-covered, and beneath are the smooth sea-ice covered waters of the bay.

freeze–thaw and other processes produce a wide range of unique phenomena, including frost wedges, patterned ground, flow deposits, rock glaciers and scree (or talus) slopes. The Arctic islands provide classic sites for the study of these phenomena.

Permafrost

Permafrost is ground that has remained frozen continually for up to thousands of years and it affects not only soil and loose, superficial deposits, but also the bedrock beneath. In summer, the upper part of the permafrost thaws out to give a so-called active layer normally ranging in depth from about a half to two metres thick. Since drainage is poor, the ground becomes very marshy and difficult to walk across. Building and engineering work is faced with serious difficulties, as disturbance of

Segregation of ice from sediment by repeated freezing and thawing is a common process on river plains, raising the overlying material into a mound. In this example from Engelskbukta, northwest Spitsbergen, the mound is collapsing as the ice core, now exposed, melts.

the ground can destabilise the soil. Furthermore, the active layer is particularly susceptible to severe erosion if it is disturbed, for example by the passage of vehicles. Even where no erosion occurs, unsightly tracks are left that may take many decades to heal.

Overcoming construction difficulties associated with permafrost has given rise to novel concepts. Commonly, buildings are constructed on wooden piles up to a few metres above ground level, in order to prevent the transfer of heat to the ground. Similarly, services such as water and sewerage need to be kept above ground level in heavily insulated, heated tubes or boxed conduits. Heated buildings, resting directly on permafrost, are likely to become structurally unstable in the space of a few years. Many older buildings in Arctic settlements can be seen to lean in response to disturbance and flowage of the active layer.

Permafrost is commonly divided into three types: continuous, discontinuous

Buildings constructed in areas affected by permafrost need to be insulated from the ground. Tried and tested techniques include placing the building on wooden piles. This photograph shows a corner of the SAS/Radison Hotel in Longyearbyen, Svalbard. Although the area between the piles is boarded off, it is maintained at air temperature as the building above is fully insulated.

Most services, such as water and sewerage, need to be insulated from both permafrost and sub-freezing winter temperatures. Because of this, these services are commonly placed on wooden stilts above ground level. This example, near the school in Longyearbyen, shows new, thickly insulated plastic piping replacing the old metal conduit. Many such services have a heating element running through them. Note the typical bridge across the pipes, used mainly as a route for snow-scooters in winter.

and sporadic. Continuous permafrost is the dominant type in all the Arctic islands, except southern Greenland. There are even zones of permafrost under the sea, especially off the North Siberian and North American mainland coasts. This submarine permafrost is probably a relict feature from times when sea levels were lower during cold, full-glacial conditions, when glaciers and ice sheets grew to cover more extensive areas than they do today. Locations such as the shallow Laptev and East Siberian seas provide examples of submarine permafrost. Intensive research is being undertaken on offshore permafrost because of the difficulties it poses for oil and gas exploration.

The depth of permafrost is extremely variable, even within a small area. Depth values have been obtained from several places by drilling. Examples from the

Old buildings of the Store Norsk Spitsbergen Kullcompani, now demolished, in Longyearbyen, Spitsbergen. One of the authors stayed in these buildings in 1983, noticing the pronounced slope of the floors in some areas.

Arctic islands include the following:

Alert, Ellesmere Island	c. 1000 m
Qaanaaq (Thule), North Greenland	518 m
Svalbard	241–305 m

Beneath thick glaciers, however, there may be little or no permafrost because the ice cover provides a blanket that insulates the ground from the cold climate, and allows heat from the interior of the Earth (geothermal heat) to warm the base of the glacier.

Permanently frozen ground can also act as a form of 'deep freeze' for animal remains. The well-preserved remains of the last mammoths have been extracted

A mammoth tusk and molar, recently melted out of permafrost in the bank of the Neizvestnaya River on Wrangel Island (Photo: J. Karhu).

from permafrost areas in the Russian Arctic, on Wrangel Island for example, often where summer melt-streams have eroded their bed and banks. The permafrost has preserved both the hair and internal organs of the mammoths, along with their characteristic and durable curved tusks.

Ground-ice

Closely associated with permafrost are various types of ice body. Ice invariably exists in the pore space in soils and sediments, or in fractures and other weaknesses in bedrock. It also occurs, more strikingly, as segregated lenses or in V-shaped wedges extending down from the surface. In addition, other types of ground-ice, such as river ice or glacier ice that have become buried by sediment, appear occasionally.

Perhaps the most dramatic manifestations of ground-ice are conical hills that may attain a height of 10 m, known as pingos. A pingo forms by injection and freezing of ground water, which is confined by frozen ground around the water body. This process causes uplift of the overlying unconsolidated deposits. Pingos are commonest in low-lying areas, particularly in river valleys, and numerous examples have been described, notably from East Greenland and Spitsbergen. Pingos are sometimes referred to as frost boils. Occasionally they 'burst', as is evident from the water-filled craters at their summits.

Another form of ground-ice, which lies just below the surface, is composed of columnar crystals, and is formed when the ground is subjected to repeated freezing and thawing, especially where water is in abundant supply. Upstanding ice crystals may reach half a metre in length in these bodies, and often underlie moss-mats or thin debris layers.

Sheets of ice up to a few metres thick, resting on the land surface, are called *Aufeis* or *naled*. They commonly develop on the floors of river valleys as a result of freezing of running water in the autumn. Glaciers often continue to maintain a flow of meltwater, even when autumn air temperatures have fallen well below the freezing point. This water freezes when it flows out of the channels within the glacier and onto the valley floor beyond. *Aufeis* comprises sheets of vertically orientated ice crystals in the form of 'needles' or 'candles'. Although the ice layer is typically a metre or so thick, under exceptionally favourable conditions sheets 3–4 m thick may develop. Aufeis may persist well into, or even throughout, the summer season. In some areas, it may be buried and preserved under glacial outwash debris.

Map of the distribution of different kinds of permafrost in the Arctic.

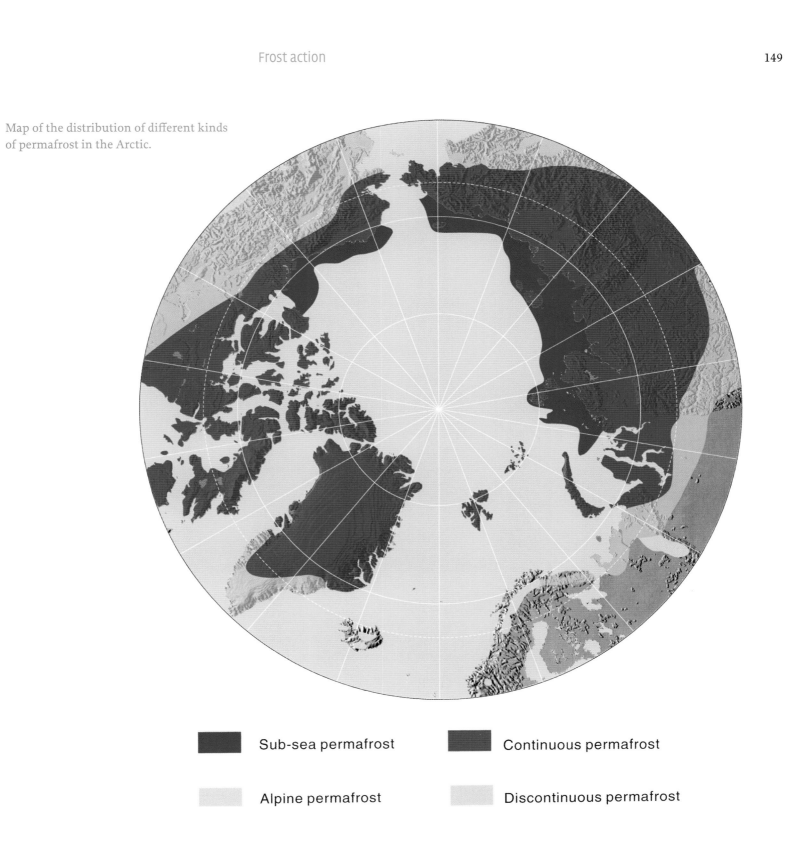

Sub-sea permafrost

Continuous permafrost

Alpine permafrost

Discontinuous permafrost

Conical hills, up to 10 m high and with a core of ground-ice, are known as pingos. They are best developed on braided-river plains where there is an abundant supply of water, as here in Reindalen, central Spitsbergen (Photo: Neil Glasser).

Patterned ground

Freeze–thaw processes in the active layer produce a fascinating array of geometrical forms, referred to as patterned ground. The patterns result from frost-sorting and cracking, and include circles, polygons and stripes. Some types are sorted, whereby larger stones are pushed to the edges of each feature, leaving a central muddy area that is very soft early in the summer season. Stone circles and polygons range from one to several metres in diameter, whereas stone stripes are a few centimetres to a metre across. Circles and polygons occur on flat ground, and are best developed on bare stony raised beaches, especially those that escaped the last phase of glaciation about 10,000 years ago. On gentle slopes, the circles and polygons become elon-

A sheet of *Aufeis* collapsing over a tunnel cut by a stream in the area in front of the glacier Pedersenbreen in northwest Spitsbergen. *Aufeis* typically has a vertical candle-like crystal structure, and forms in the autumn from glacier-discharges after the temperature has fallen below freezing.

gated and grade into stone stripes on steeper slopes of loose debris. On steeper slopes, freeze–thaw processes, combined with removal of fine sediments by running water during spring, sort the sediment to give so-called stone streams. Non-sorted circles and polygons are the result of frost-cracking and are defined at the surface by narrow fractures in the soil and vegetation. These fractures disguise the presence of wedge-shaped lenses of ice up to half a metre wide and several metres deep below the surface. All of these forms are gradational with one another. The larger forms, developed over thousands of years, provide complex ecological niches for vegetation growth. Active stone stripes may come and go within one summer season.

Downslope creep of saturated ground

On gentle slopes, particular those that are partially vegetated, downslope creep of saturated soil and superficial debris takes place – a process known as solifluction. Old glacial deposits, which contain a large amount of clay that can readily hold

The geometric pattern of frost cracks shows clearly in this aerial view of the valley of Gipsdalen in Spitsbergen. Although the photograph is taken in spring, before the melt season has begun, wind has blown much of the ground-surface free of snow.

water, are particularly susceptible to creep. Rates of movement are irregular, but on average amount to a few centimetres a year. This irregular movement produces lobate patterns, with banks up to a metre thick often retarded by vegetation growth. Solifluction is a process that can take place on slopes as gentle as 2 degrees to as steep as 20 degrees.

Weathering of bedrock

Frost-shattering is an important process throughout the Arctic islands, and leads to the development of craggy slopes and rocky towers, whilst the slopes below are mantled with scree or talus. Frost-shattering involves the expansion of ice occupying cracks in bedrock as water freezes. The little-known phenomenon of thermal

Frost-sorting in areas where snow-melt is trapped and mobilises mud-rich glacial sediment, yields stone polygons as here near the summit of the highest hill on Blomstrandøya in northwestern Spitsbergen. The stone ridges make for firm walking, but boots can be overtopped if they sink into the fine material in the middle of the polygon.

Frost-sorting on level stony ground produces stone circles, such as these in the Murchisonfjorden area of Nordaustlandet, Svalbard. These circles have an external diameter of approximately 1 m.

Frost-sorting on slopes results in stone stripes. In this example, in the Murchisonfjorden area of Nordaustlandet, the scree slope provides a steady supply of angular rock fragments, which are then sorted as the material is carried down to the wetter more muddy ground below.

stress fatigue of bedrock may also take place in circumstances when there is little water available. As radiation from the sun is absorbed by rock surfaces, when the air temperature is very low, a strong thermal gradient develops between the shady and sunny sides. This generates strong internal stresses, which cause the rock to fracture. Another poorly researched phenomenon is biological weathering by bacteria and algae. Algae, for example, expand during wetting episodes, and at the same time induce chemical changes, facilitating the flaking of bedrock.

Scree slopes below the cliffs are composed of angular blocks of rock whose size reflects the structure of the bedrock. Scree slopes typically show grading of material from fine gravel at the top to large boulders at the bottom. Reworking of material by short-lived streams, formed beneath the winter snow cover in spring, can modify scree slopes considerably. Although many scree slopes are clearly actively forming today, the rates of recession of the cliff faces above are surprisingly low. Indeed, some scree slopes in the more maritime parts of the Arctic islands, for example the west coast of Spitsbergen, have become partially stabilised; birds

Frost-sorting and meltwater runoff combine to produce a 'stone stream' of angular cobbles and boulders, a common feature of moderately inclined slopes. Here we are looking downslope at a stone stream on Wolf Mountain, Axel Heiberg Island, Canada.

nesting in the cliffs above supply guano to the slopes below, encouraging mosses, lichens and grasses to grow.

Scree slopes commonly fan out from gullies where snow accumulates preferentially. The melting snowpack provides water that flows into spaces between boulders in the scree. If sufficient 'interstitial' ice accumulates the whole mass can deform, and a rock glacier develops. Where best developed, rock glaciers form lobate features, up to 100 m or more high, and 500 m long and wide, extending out from the slope. Flow may be rotational (i.e. outwards and upwards) and, commonly, a hollow is created at the back of the feature against the main slope. Here ponds may develop, enhancing the supply of water to form interstitial ice. Lobate rock glaciers are especially common in East Greenland and western Spitsbergen. Some of the more active rock glaciers flow downslope like normal glaciers, producing a very irregular surface topography. A further type of rock glacier occurs occasionally, simply by burial of a 'normal' glacier by a continuous cover of frost-shattered debris from the valley walls.

A rock glacier, with a steep frontal rampart and depression behind it, sits at the edge of a sea-ice covered fjord in eastern Ellesmere Island. Such features are formed as a result of deformation of interstitial ice, derived from water that percolates into the scree slope from melting snow. Inside the lobes, small lakes commonly form. Frost-shattered scree cones also cover much of the steep sidewall of the fjord.

Scree slopes are often modified by water when the winter snow pack that covers them begins to melt. Fast-flowing streams under the snow are confined in channels, and distinctive levées of sediment are developed along their flanks, as here in Bolterdalen, central Spitsbergen.

The effects of frost weathering are illustrated in this view of the plateau edge above Kap Weber in East Greenland. On the left, crags are yielding a steady supply of scree material, whilst towers and pinnacles are developing on the right as the plateau edge disintegrates. The scree slopes fall in a continuous sweep down to the fjord waters, hidden beneath the fog bank, over 1,000 m below.

Another landscape feature that is characteristic of frost action, which occurs on high-level plateau areas, is blockfields which are also commonly known by their German name, *Felsenmeer*. Blockfields result from the breakdown of bedrock along joints, bedding planes and other internal structures. They consist of blocks up to several metres across that are more or less *in situ*, but have tilted as the bedrock disintegrates. Slow movement of such blocks may take place on gentle slopes. Bedrock disintegration, whilst largely complete in many areas, may leave isolated masses of bedrock still intact. Such upstanding masses, several metres high, are known as tors. Like most landforms of frost-shattering, they are not confined to the Arctic, and many examples occur in cool temperate upland regions elsewhere.

Vulnerability of permafrost areas

We have noted the direct influence of humans on land underlain by permafrost, and the damage that may be caused by careless construction and motorised

Talus cones and stone chutes reach to sea level in the steep-walled fjord of Tinayrebukta, an arm of the Krossfjorden system in northwest Spitsbergen. Note that some of the talus has vegetation growth, indicating greater stability, whereas lighter-coloured areas are more active.

transport. However, there are potentially more significant processes at work that could have global consequences. In a warming world, permafrost may melt, causing wholesale collapse of ground above ice bodies, releasing large amounts of freshwater into the surrounding oceans. Permafrost is associated with a particular type of vegetation, tundra, and as this becomes prone to destruction, large amounts of methane could be released. Since methane is an extremely potent greenhouse gas, its release by the melting of permafrost terrain could contribute significantly to global warming. These effects are poorly understood and, of course, difficult to monitor.

A wide range of slope processes are evident in this picture of the coast of Traill Ø, East Greenland, where it borders Kong Oscar Fjord. A rock glacier dominates the middle of the picture, in this case forming a broad uneven tongue extending into the sea. In the background, scree slopes and small alluvial fans are developing below the crumbling cliff. In the foreground the smoother, braided surface of a delta is supplied with sediment from a small river. Coastal processes are reworking the sediment to form beach ridges.

A lobate rock glacier in Brogetdal, East Greenland, developed on actively forming scree slopes. A wisp of cloud floats over the lobe itself.

The frost-shattered peaks and ridges of western Spitsbergen provide sources of debris to the surrounding glaciers. This scene is of Oscar II Land, with the glacier Aavatsmarkbreen in the background.

Chapter 7

Arctic coasts, rivers and lakes

Most visitors to the Arctic tend to remain within easy reach of the coast and only scientific groups, working on the ice masses, and trekking or skiing expeditions venture far inland. Nevertheless, the coastal zone with its shorelines, rivers and lakes provides the visitor with much of interest. Sitting on an Arctic beach under the Midnight Sun, as stranded icebergs wallow and crackle in the shallow waters, while seabirds dive in and out of the sea, and at the same time keeping a watchful eye out for polar bears, can be a memorable experience.

Glaciers at the coast

Islands, by their nature, have extensive lengths of coastline. The action of ice, in the form of both glaciers and sea ice, has left a particular imprint on the coasts of the Arctic islands. A significant proportion of the Arctic coastline is made up of ice cliffs, marking the terminus of glaciers and calving icebergs into the sea. In addition, the distribution of rivers and lakes on the Arctic islands is also influenced strongly by both the present and past impact of glaciers.

As well as providing a source of icebergs, glaciers produce large volumes of meltwater during summer. This meltwater reaches the sea in streams that flow through the glacier, ultimately reaching the bed and discharging at the base of ice cliffs resting on the sea floor. These subglacial streams, carrying large quantities of debris to the sea, are identified by plumes of muddy, turbid water flowing away from the ice front. The streams flow from tunnels at the base of the glacier and

The coastal zones of Arctic areas such as Kongsfjorden, northwest Spitsbergen, as depicted here, are full of interest. Here, a sizeable iceberg is drifting past the four onlookers as bergy bits litter the gravel shore amongst their feet.

A sediment-laden meltwater stream flows into the waters of Kongsfjorden, northwest Spitsbergen. The meltwater spreads out on the sea surface because it is less dense than the saline marine water. Small, isolated lakes are also present on the peninsula, and sea ice has been piled up against its shore by the wind.

immediately rise to the surface of the sea. The meltwater rises because it is fresh rather than saline and is, therefore, much less dense than sea water, despite its relatively high load of suspended sediment. Larger debris, in the form of pebbles and boulders, is not transported beyond the point at which subglacial streams enter the sea, and this material is deposited as an underwater delta or debris-fan close to the base of the ice cliffs, where it is mixed with sediment derived directly from the ice.

Fjords

Even where glaciers are no longer present at the coast, they have left traces of their power as agents of landscape erosion. Fjords are deep troughs with steep rock walls and flat bottoms that have been inundated by the sea. Most fjords have concave slopes rising into cliffs above the waterline and, in cross-section, could be described as having a parabolic profile. Some fjords are vertical-sided, with a clearly U-shaped cross-section. Many fjords still influenced by glaciers are located in the Arctic islands and elsewhere, including Alaska and Patagonia. Fjords that are now ice-free are typified by the west coast of Norway, from where the term is derived, but many coastal mountain areas that have been affected by the former presence of glaciers and ice sheets also have fjords; the sea lochs of Scotland are an example.

Most of the larger Arctic islands contain fjords, cut into upland areas, and often following the geological structure of the bedrock, such as faults. Fjords are probably the most distinctive landscape feature of Arctic coasts, together with the glaciers that characteristically end at their heads. Many fjords are straight, but others are sinuous and have many side branches. They can extend hundreds of metres below sea level, and range from a few kilometres to over 100 km in length. The world's largest fjord is Scoresby Sund, with its northern branch of Nordvestfjord, in East Greenland. It stretches inland for over 300 km and reaches water depths of over 1,500 m.

Fjords are among the most dramatic effects of glaciation, extending, as with Kejser Franz Josef Fjord in East Greenland, more than 100 km inland. This view shows icebergs drifting from the distant glaciers at the head of the fjord, as well as a tributary glacial valley (upper right) and a delta (right foreground). The solitary peak in the middle of the view is Teufelschloss (meaning Devil's Castle), over 1,300 m high.

Icebergs and sea ice litter the calm waters of Kangerlussuaq Fjord in southeast Greenland. The fjord is about 70 km long and up to about 1,000 m deep.

Where ships, fitted with depth-sounding equipment, have surveyed fjord waters, shallows known as sills are commonly found at fjord mouths. Sills may be made of rock or glacier-deposited moraine, or both. They mark the position to which glaciers extended in the past. The steep fjord walls contain rock faces and ledges that provide a home for seabirds such as auks and gulls during the nesting season. These rock walls shatter as a result of frost action, and form huge cones of bouldery talus or scree.

Almost the entire coast of Greenland is dominated by fjords. In addition to Scoresby Sund, several fjords exceed 100 km in length, including Kejser Franz Josef Fjord, Kong Oscar Fjord and Kangerlussuaq on the island's east coast. All these fjords were carved out by fast-flowing outlet glaciers emanating from the Greenland Ice Sheet, and the process is still continuing at their heads.

The granite cliffs of the Meta Incognita
Peninsula line the southeastern shores of
Frobisher Bay, the southernmost fjord in
Baffin Island.

Glacial rivers carry a large amount of sediment to the coast. The finer materials may be subject to reworking by other processes such as wind. Here we see an inland sand-dune system derived from the fine sediment in glacial meltwater at Albert Heim Bjerg in East Greenland.

In the Canadian Arctic, the highlands on the east coast of the 1,500 km-long Baffin Island contain over 200 fjords, and the coasts of Ellesmere and Axel Heiberg islands, further to the north, also provide numerous examples. By contrast, fjords are not present in the lowland coastline that makes up much of the west side of Baffin Island.

In the Eurasian Arctic, the island of Spitsbergen contains many beautiful, but relatively short fjords, the largest being on its west and north coasts. Hornsund, Bellsund, Isfjorden and Kongsfjorden in the west and Woodfjorden and Wijdefjorden in the north, effectively divide the island into a series of discrete regions. In the Russian Arctic, too, parts of the coasts of Severnaya Zemlya and Franz Josef Land are made up of fjords, whereas fjords are not well developed in the low-lying New Siberian Islands, which contain no glaciers today. The large northern island of Novaya Zemlya is also heavily dissected by fjords although, unfortunately, some of these deep waters have been used as dumping grounds for radioactive waste including, it is thought, nuclear reactors from the Russian fleets.

The dynamic nature of Arctic coasts is illustrated by this photograph of Engelskbukta in northwest Spitsbergen. The glacier at the head, Comfortlessbreen, supplies a large volume of sediment directly or in suspension, whilst the river from another glacier in the foreground delivers a large amount of fluvial material. This material is reworked by waves and coastal currents, producing beach bars and spits, behind which are muddy lagoons.

Arctic shorelines

The coastline of the Arctic islands receives large quantities of the debris that is eroded by glaciers. Glaciers pick up this debris and transport it directly to the coast within the ice itself or through subglacial melt streams, or indirectly from glacier-fed rivers. Meltwater, and the sediment it carries, flows mainly during the summer melt season, and discharge and sediment output is minimal during winter. Sandy sediments deposited in river valleys can be remobilised into sand dunes by the wind. The debris delivered to the sea from glaciers and rivers fed by glacier and snowmelt is then transported by currents. Much of this material is deposited as beaches of either sand or gravel. Boulders sometimes litter these beaches, and were probably transported there by icebergs or sea ice, or were eroded from adjacent moraines. Long spits and bars, often containing material that varies gradually from coarse to finer particles along their length, are typical of many Arctic coasts where longshore currents predominate. Shorelines consisting primarily of boulders are

Close-up view of the shoreline of Engelskbukta, shown in the previous picture, illustrating the shingle beach. The pits were formed by the melting out of sea ice floes or small icebergs after burial by sediment, and are a common feature of Arctic shorelines.

common. The building of beach ridges is assisted by the presence of sea ice during winter. Winds can push the sea ice up the face of the beach and this mechanical action helps to build the ridges of boulders that are typical of many Arctic beaches, whilst wave action winnows out and removes the finer material.

On many beaches in Svalbard you can observe driftwood in the form of logs, sometimes 10 m or more in length. The logs are derived from Siberian rivers, washed into the Arctic Ocean during spring floods. They then drift across the ocean in the grip of the pack ice. Unfortunately, Arctic beaches are not immune from that curse of beaches in temperate latitudes, human litter. Plastic bottles and fishery debris are not uncommon.

Many coastal areas of Svalbard consist of a low plain known as a 'strandflat'. These plains are present just above sea level and may be up to 5 km wide. The flat surfaces commonly extend offshore, where they form low-level rocky islands or skerries, which are a hazard to inshore navigation. The origin of strandflats is uncertain, but some scientists believe that they are a product of weathering above sea level, followed by erosion by the sea. Such processes would have been repeated several times during glacial–interglacial cycles as a result of repeated raising and lowering of sea level.

Sets of raised beaches can also be traced above the modern shoreline on many Arctic islands. When the ice sheets, which covered both high and mid-latitudes during the last full-glacial period about 18,000 years ago, decayed, their weight was removed from the Earth's crust. With the release of this load, the crust rebounded upwards, bringing with it the beaches formed at sea level soon after deglaciation. Crustal rebound continues in the Arctic islands today, the amount of uplift being dependent on the thickness of the ice that covered different parts of the Arctic, and the timing of deglaciation which varied across the region from 10,000 to 6,000

Flat coastal plains, criss-crossed by glacial rivers, are known as strandflats, and are common features of western and northern Spitsbergen. This view shows a 3 km-wide strandflat along the coast of Oscar II Land to the south of Engelskbukta.

Raised beaches are a feature of areas that have 'rebounded' following the removal of the ice-load during deglaciation. The bay (upper left) has a regular set of raised beaches, which appear, coincidentally, to line up with the tilted rock strata in the foreground. This view is of the southern entrance to Van Keulenfjorden in south-central Spitsbergen.

years ago. This uplift has taken with it the beaches formed at different times over the past 10,000 years or so, and has produced the 'staircase' appearance so typical of many Arctic shorelines. The timing of formation of these raised beaches can be calculated accurately using radiocarbon dating of organic carbon, often obtained from the bones of whales, shells and the remains of driftwood that are incorporated into beach material.

The amount of post-glacial uplift is extremely variable throughout the Arctic islands. In Greenland, typical uplift values for Peary Land in the north, Mestersvig in the east and the west coast are of the order of 120 m. Over the same period in Svalbard, we find values of around 60 m for Murchisonfjorden on Nordaustlandet and in Billefjorden, central Spitsbergen. Cape Hooper on Baffin Island has risen some 42 m since deglaciation, 9,000 years ago. Uplift curves are exponential, that is, they show decelerating uplift through time, so the bulk of uplift is achieved within a few thousand years of deglaciation. Not all areas show uplift, however: those

Driftwood of tree-trunk proportions litters many Svalbard beaches. The logs in this photograph lie on a beach on Nordaustlandet bordering the strait of Hinlopenstretet, and are derived originally from Siberian rivers, but transported here in the pack ice of the Arctic Ocean.

areas close to the ice-sheet margin, such as the northwestern tip of Spitsbergen, have actually experienced a relative sea-level rise.

Cliffed shorelines are common coastal features, especially along fjords. Those facing the open sea rarely attain substantial heights, but notable exceptions are found on Bjørnøya (Bear Island), the small island that lies roughly halfway between Spitsbergen and mainland Norway.

Braided river systems

Large rivers, with numerous branching channels, known as braids, separated by 'bars' of sand and gravel which are constantly on the move, are typical of the landscape of the Arctic islands. These rivers are fed by meltwater, both from snow and from those glaciers that terminate on land. In Spitsbergen and on Axel Heiberg Island, for example, extensive valleys, often tens of kilometres in length, are

Bear Island (Bjørnøya) is a small remote island in the Barents Sea, and displays a fine set of 400 m-high cliffs and sea stacks, rising up into the clouds that almost perpetually shroud this area in summer.

During the early summer, when snowmelt is at its peak, and the glaciers are beginning to melt, braided rivers become very full and often impassible. This is the river that emerges from the large outlet glacier Aavatsmarkbreen, on the west coast of Spitsbergen, discharging into the sea opposite the heavily ice-covered island of Prins Karls Forland.

Large river channels in the valley of Reindalen in Spitsbergen. Streams enter the valley from both sides, fed from the melting of glaciers in the mountains above. The intricate pattern of channels is known as braiding, where streams branch and rejoin repeatedly. Braided rivers are common in the Arctic, and are formed where meltwater flow is very variable on a daily to annual basis.

A common feature of High Arctic glaciers is the continued slow release of stored meltwater onto the braided river plain in front after the winter freeze has begun. As a result, thick masses of candle ice called *Aufeis* build up during the winter, often to a depth of 2 m, as here at Midtre Lovenbreen in northwest Spitsbergen. This ice may survive late into the summer, or even through to the following winter.

dominated by networks of fast-flowing rivers during summer. The peak in river discharge is related to snowmelt and usually occurs in late June or early July, depending on latitude and the weather in any given year. At such times, much of the valley floor can be inundated by water. These rivers can also be cut deeply into the landscape through which they flow, even forming steep-walled canyons. For example, the landscape of western Devon Island, in Arctic Canada, is heavily dissected by networks of rivers set within deep valleys. Elsewhere, flat terraces are left high and dry, providing ideal habitats for a variety of Arctic plants and animals. The postglacial rebound of the land surface, responsible for the formation of raised beaches, is also the source of energy that leads to downcutting by Arctic rivers. Many Arctic rivers are turbid from suspended sediment. This sediment comes from the readily erodible valley sides or from the glaciers.

Where braided rivers enter the sea, deltas usually form, because the river

Braided rivers contain channels that are actively changing right across the valley floor. As they enter the sea the sediment is largely deposited and a delta builds out from the coast. Often these deltas form in embayments previously carved out by the ice. This example shows a delta in northern Jameson Land, looking out across the mouth of Kong Oscar Fjord in East Greenland.

Crossing braided rivers requires a bit of a knack, as the straight line approach will invariably mean wet legs. The trick is to link mid-stream bars together via wide, fast-flowing shallow channel crossings. This is the sediment-laden braided river from Midre Lovénbreen in northwest Spitsbergen, a rather small example by Arctic standards, but nevertheless still a significant obstacle to travel on foot.

deposits much of its debris load when it leaves the confines of its channel. Debris reaching the delta front is immediately subject to further transport by wave action or long-shore currents, forming beach ridges. Behind these ridges, muddy lagoonal areas, favoured by wading birds, are commonly formed.

A unique feature of some rivers that flow from glaciers is the presence of extensive sheets of water ice, known as icings or *Aufeis*, on the braid plain. Where the glacier bed reaches the melting point, water can be stored at the bed and released slowly during the winter. As this water emerges from the glacier, it begins to freeze and sheets of ice with a candle-like structure accumulate to depths of up to two metres or so. These sheets of ice can survive well into the summer season, and sometimes into the next winter. Occasionally, this ice is buried by sediment to a depth of several metres, and in such circumstances can probably survive for hundreds of years.

Arctic rivers provide a significant barrier to travel in the interior of Arctic islands during the summer melt-season. Even small streams must be crossed with great care because of near-zero water temperatures, the typically rocky and unstable nature of their beds, and the lack of visibility through the turbid water. Once above knee-height, these rivers are often too dangerous to cross. It is for this reason

Winter icings or *Aufeis* constrain the flow of this river of glacial and snow meltwater that flows in a small valley between the large lateral moraine of Kongsvegen (left) and the valley side, northwest Spitsbergen.

The surface of two lakes in Signedalen, northwest Spitsbergen, still partly ice-covered even in late July. Lakes such as these are fed from early season snowmelt, together with meltwater from glaciers throughout the summer. Some of these lakes contain fish, such as Arctic char. In the background are the mountains of the neighbouring island of Prins Karls Forland.

that much of the early exploration of the Canadian and European Arctic islands took place in spring, when daylight approaches 24 hours, river-flow has yet to begin, and smooth shorefast sea ice remains around the coast. For example, the islands of Severnaya Zemlya were first explored only in 1930 by the spring sledging expedition led by Ushakov. Even today, the Danish Sirius Patrol still use dogs to sledge along the east coast of Greenland during winter.

Lakes and their environmental record

In low-lying parts of some Arctic islands, such as Prince of Wales and Steffansson islands in the Canadian Arctic, many lakes are found. The rock basins in which

A fresh-water lake on Ymer Ø, East Greenland, fed only by snowmelt, occupies a rock basin carved out by glaciers which flowed from the mainland peaks in the background. The clear fresh water provides an ideal habitat for Arctic char, while the shores bear a diverse flora and fauna.

these lakes form have been produced by the scouring action of the ice sheets which last covered these areas about 18,000 years ago. The lakes are similar in shape and origin to those found today in northern Minnesota and the Finger Lakes region of New York State. The surface of these Arctic lakes remains frozen for much of the year. The breakup of lake ice is typically later than that of ice in the adjacent seas. This is for two reasons. First, lake waters are fresh, rather than saline, and therefore have a freezing point of 0 °C, rather than the −1.8 °C of sea water. Secondly, lakes generally have weak currents and no tides, which act to break up sea ice mechanically as it thins during early summer.

The sediments deposited in Arctic lakes contain a good record of the changes in the Northern climate over the past 10,000 years or so. Cores from Lake Hazen in

Glacier-dammed lakes are a common feature of Arctic glaciers. Here a branch of an outlet glacier of the Greenland Ice Sheet in East Greenland, called Dödehunds Gletscher (Dead Dog Glacier) has dammed a small frozen lake. Tabular icebergs are breaking off from the floating glacier tongue.

northern Ellesmere Island, for example, give a picture of the changing environment in the Canadian High Arctic over the past few thousand years. Sediment is delivered by streams flowing into the lakes mainly during summer, usually fed from snow and glacier melt. In winter, stream-flow ceases and only the very fine clays left in the lake waters settle to the lake floor at this time. Thus, each year a summer layer of sand and silt is laid down, followed each winter by a thin layer of finer clay. In some Arctic lakes these rhythmically-deposited pairs of layers, known as varves, record variations in climate through the changing thickness of the summer and winter layers over periods of hundreds and sometimes even several thousands of years. Lakes throughout the Arctic islands also contain Arctic char, a member of the salmon family, and often provide good fishing in high summer.

The fringes of the lakes are often boggy and home to a relatively rich flora,

Glacier-dammed lakes commonly drain every summer in catastrophic fashion. Here, Between Lake, sandwiched between two converging glaciers on Axel Heiberg Island, the White on the left and Thompson on the right is in the process of draining, and the high-water mark is clearly visible at the level of the top of the waterfall.

which in turn provides sustenance to a variety of insects, birds and mammals. Although these lakes are relatively permanent features of the landscape, there is another type of lake which only forms sporadically, or sometimes seasonally – those dammed by glaciers. Glacier-dammed lakes form where side valleys are blocked by ice, or in the 'V' where two glaciers converge, or simply along the ice margins. Typically these lakes fill up during the summer until a sufficient head of water causes the ice to lift, thus releasing the water as an unexpected flood. As the lake fills, icebergs calve into it, and these are left stranded when the lake empties. Some lakes do not follow the seasonal pattern of filling and draining. They can remain empty if the conduits draining them through the ice remain open, as in relatively stagnant glaciers, or they can stay full and simply spill out across the ice surface if the flow of ice is sufficient to keep any cavities closed. Typically, glacier-

Commercial exploitation of whales in
the Arctic during the seventeenth to
nineteenth centuries has rendered them
largely extinct, and the remains litter the
beaches of Svalbard to this day. This
example of part of a whale skeleton lies
on the southern shore of Van Keulenfjorden
in south-central Spitsbergen.

dammed lakes are turbid, and this characteristic, together with their ephemeral nature, means that they can support little life.

Human activity on the Arctic coastline

The coastlines of the Arctic islands, and occasional lake shores, have also been home to those who have lived in the North, from the earliest Inuit hunters, through whalers, hunters and trappers, to the few modern settlements that are found in the high Arctic. It is around the coastline, in particular, that much of the animal life of the Arctic is concentrated during summer, when the winter sea-ice cover has broken up, and it is this that attracted human civilisation here too. The sites of early human occupation are marked, for example, by the stone rings of Inuit summer camps that can be found in the Canadian Arctic islands. In the Eurasian Arctic, the islands have no indigenous peoples, and the earliest signs of human presence are often recorded in the remains of whaling camps, one of the best known being the seventeenth century site at Smeerenbergfjorden in north-

An aerial view of Grise Fiord, the most northerly civilian settlement in the Canadian Arctic, located in southern Ellesmere Island. Beyond the mainly Inuit-occupied houses, the tracks of many snowmobiles run out over the sea ice that covers the waters of Jones Sound during winter and spring.

western Spitsbergen. There are also a number of graves scattered around the shorelines of the Arctic islands, often associated with whaling and trapping activity. Coffins, buried at shallow depths because of large beach boulders and permafrost, have in some places been brought to the surface by the action of ground freezing and thawing or as a result of coastal erosion.

Recent human activity has also been concentrated along the Arctic coastline. This is because the coast is both accessible to shipping and contains some areas of relatively flat land where airstrips have been built. Fuel is the lifeblood of modern Arctic settlements, and it is only through summer visits by ships that sufficient quantities can be brought in from the South. Most Arctic bases, both military and civilian, are located at the coast and are resupplied annually by ship. Their freshwater supply is often obtained from small lakes nearby. The northernmost settlements in the Canadian Arctic, at Alert and Eureka on Ellesmere Island, are both in coastal locations with access for shipping and airstrips where aircraft can land. Alert, at 82.5° N, has its origins as a military base during the Cold War years, whereas Eureka (80° N) was established as a weather station in 1947. In the Russian

A refuelling point for Russian helicopters is located close to the shore of Hayes Island in the Franz Josef Land archipelago. The broken surface beyond the helicopter is formed from small icebergs and old sea ice embedded within the frozen sea surface formed during the current winter.

The coastal coal-mining settlement of Svea
provides the foreground to the glaciers and
mountains of central Spitsbergen. Svea is
on the shores of inner Van Mijenfjorden,
and the tidewater glacier Paulabreen can
be seen in the distance at the head of
Rindersbukta. The fjord is usually
navigable from about mid-July until
September.

Arctic, strategic and logistic considerations have also meant that settlements have
grown up since World War II in coastal locations. Military and meteorological
bases such as Sredniy in Severnaya Zemlya and that on Hayes Island in Franz Josef
Land were established around the Arctic coastline. In Svalbard, the Norwegian settlements of Longyearbyen, Ny-Ålesund and Svea, and the Russian towns of
Barentsburg and Pyramiden, were set up to exploit coal reserves. All are located on
the coast, and supplies are brought in by ship during the brief Arctic summer.

The rocky coastline of Wrangel Island,
north of Siberia (Photo: M. Saarnisto)

Chapter 8   Animals and plants

Introduction

The Arctic islands are home for diverse and specially adapted plant and animal populations. The polar bear, a once-endangered species that is now protected, epitomises the Arctic, spending much of its life on polar pack ice, feeding on seals that inhabit the waters beneath in large numbers, but also frequenting the coastal fringes. Whales, walrus and some seal species, although also now largely protected, are only slowly recovering from over-exploitation by humans. On the islands themselves, the natural habitats of musk ox, lemming, fox, hare, caribou and reindeer extend to the northern limits of the land. Birds migrate in vast numbers to breed in the Arctic in summer, but only a few species remain there all year round. One species, the Arctic tern, breeds in the High Arctic but undertakes a phenomenal round-trip to the Antarctic to take advantage of the Southern Hemisphere summer when the Arctic is in darkness, and food is in short supply. This small bird sees the greatest proportion of daylight in its life of any living thing.

The Arctic islands support a beautiful flora – at least at low altitude. As the winter snow melts, many plants burst into flower. Delicate mauves and yellows light up the bare tundra, as species such as purple saxifrage, moss campion, arctic poppy and mountain avens flower. The flowering period is less than two months, but in this brief spell the Arctic is at its best. Ground-hugging birch and dwarf willow grow well in sheltered parts of the Arctic, a few reaching half a metre in height. Towards their more northern limits, such as Svalbard, willows are only a couple of centimetres high. Insects breed in profusion during the short summer, especially mosquitoes. Mosquitoes are a major irritation to humans and other

The magnificent Arctic landscape provides a fine backdrop to observing the activities of animals such as these reindeer (*Rangifer tarandus platyrhynchus*) near Kongsfjorden in northwest Spitsbergen. Land mammals like reindeer sometimes have a hard time locating the mosses and lichens on which they depend, especially in spring when snow cover is thickest.

Well-drained but moist tundra slopes have a rich vegetation, with several species growing side by side. Here in Spitsbergen, the yellow mountain avens (*Dryas octapetala*), purple moss campion (*Silene acaulis*) and the dwarf shrubby polar willow (*Salix polaris*) form a cushion-like carpet on the ground.

animals alike, but north of 75° their numbers decline rapidly, and in the High Arctic they largely cease to be a problem.

For plants and animals, the Arctic islands are a harsh place to live, but they are uniquely adapted to the extreme low temperatures and winter darkness. Compared with other parts of the world, the diversity of animal and plant species is low. For example, of around 3,200 mammal species world-wide, only about a dozen live on the Arctic islands, and of over 8,600 bird species, only six or seven winter and seventy or so migrate North to breed. Thus, relationships between species are relatively simple. For those organisms that have adapted successfully to the harsh environment, the Arctic provides a safe place for the fittest. However, the ecological balance between some species has been disrupted severely by humans, and only in the latter part of the twentieth century has the realisation grown that a sustainable animal population adds to the richness of our planet, and that effective management is essential to conserve vulnerable species. Although hunting and trapping for furs and other animal products continues, at least it is now on a much reduced scale, which has allowed populations of such species as polar bear, musk ox, wolf and arctic fox to recover significantly. The main threats to flora and fauna now come from indirect but potentially more severe human effects, notably atmospheric pollution dominated by greenhouse gases such as carbon dioxide, leading to climatic warming, reduction of stratospheric ozone producing a thinning or 'hole' that lets in damaging ultraviolet rays, and destruction of habitats. Another insidious threat is from airborne pollutants such as polychlorinated hydrocarbons (PCBs), which may severely affect the breeding capability of Arctic species, and the subsequent health of the young.

A suitable environment for plants and animals

Several mammal species are specially adapted to Arctic conditions, although there are related species outside the region too. Every mammal needs to generate its own heat, and this is 'expensive', so it must be well insulated to prevent heat loss. In terrestrial mammals, insulation varies according to season. Larger mammals (such as polar bear, fox and reindeer) build up fat in summer to provide reserves of energy through the winter, and grow thick coats in winter for insulation. Smaller species (such as lemmings) live in the shelter beneath the winter snow cover.

The harshness of the Arctic climate during summer is often exaggerated. Temperatures around the coast are generally above zero in the months of June, July and August and may reach as high as 15–20 °C on occasion. Furthermore, there is the advantage of 24-hour daylight throughout much of this period in the more northerly regions. The ground thaws and there is abundant moisture in many areas. These summer months allow breeding cycles to be completed effectively although quickly. In contrast, temperatures many degrees below zero, with strong winds, drifting snow and frozen ground, characterise the winter. Overall, these conditions are tolerable for a number of well-adapted plant and animal species, to the extent that in Greenland, for example, there are 500 different species of ferns and flowering plants, and probably several hundred species of insects.

Marine mammals, with their thick layers of blubber, are adapted for life in waters with a temperature that is often a degree or so below zero. Cold-blooded animals (fish and invertebrates) can live close to the freezing point but possess components in their body fluids that behave like antifreeze. Most birds are not especially adapted to Arctic conditions: they arrive only for the summer when conditions are favourable. The few resident bird species, however, have adapted to the harsh winters through, for example, the development of insulating feathers and fat layers.

Plant species diversity declines rapidly with altitude, and the ground is invariably largely bare above about 500 m. Areas vacated by glaciers over the past hundred years or so similarly have few species. It takes hundreds or even thousands of years for a well-developed flora to colonise and develop in such areas. Most plant species are native to the Arctic, surviving through previous ice ages on ice-free islands of rock known as refugia. As the ice receded, these species spread slowly over the region. Thus, broadly similar species occur on all Arctic islands. Pristine areas of the tropical and temperate climatic zones have far more species, however. We can attribute the more limited ecological diversity in the Arctic to, first, a severe climate; secondly, the newness of the environment, much of it having been covered by glacier ice until 10,000–18,000 years ago; and, thirdly, the relative uniqueness of the environment today compared with the latitudinally more southern and climatically warmer conditions of the geological past, thus limiting the available time for evolution.

Life in the marine realm

The Arctic Ocean and surrounding seas are largely ice-covered in winter and much ice remains in summer. Surface waters, therefore, are typically between 0° and −1.8 °C in the Arctic Ocean, with temperatures depressed below zero depending on the concentration of salt. A human immersed in such water would die within half an hour, and often much less. In contrast, other warm-blooded animals, such as seals, whales and polar bear, have extensive layers of fat (blubber) that reduce heat-loss to a minimum.

Plants and animals of the Arctic seas form a web of interacting communities, based on feeding relationships, known as an ecosystem. Carnivorous marine mammals are dependent on species lower in the food web by feeding on fish, which in turn feed on plankton. At the lower end of the food web, microscopic organisms grow in surface waters and under the sea ice in summer. Phytoplankton consists mostly of single-celled organisms, including diatoms and flagellates. These absorb mineral salts and carbon dioxide from the sea water and use the energy of sunlight to photosynthesise food, storing the energy as carbohydrates and lipids, to promote cell growth, movement and reproduction, while oxygen is released as a by-product.

Arctic seas are cold and ice-infested, yet are attractive to sea birds, such as these black guillemots (*Cepphus grylle*) near the tidewater glacier of Kronebreen, northwest Spitsbergen. This is especially true in summer, when the ice breaks up and marine productivity increases dramatically.

Phytoplankton, which grow only slowly in winter when there is little light, reach peak abundance in spring, when many are consumed by increasing numbers of animal plankton (zooplankton). In turn, plankton sustain fishes, sea birds and mammals. However, because the abundance of plankton only occurs for about two months of the year, many consumer species have to migrate south in winter. Following death, plankton remains settle out onto the sea floor at the end of the summer, where some become food for bottom-living (benthic) animals. The remaining material accumulates as sticky sediment called ooze, dominated by the skeletal remains of the silica-rich diatoms.

At intermediate positions in the food web are several species in the Arctic seas, including fish such as bullhead or sculpin, eel-pout and cod. The eggs of some species of polar fish hatch in shallow water and the young feed on plankton in summer, before living the remainder of their lives in deeper water. Some fishes, notably the Arctic cha (related to salmon), alternate between the sea, rivers and lakes. Char hatch in fresh water; after four or five years they descend to the sea, if there is a suitable connecting river, where they grow fat on plankton and smaller fish such as capelin. They return to lakes to breed after a summer at sea. Arctic char is a favourite delicacy of the Arctic peoples, not to mention other visitors to the Arctic, with a flavour similar to, but more subtle than, that of salmon. Interestingly, Arctic char survive in land-locked lakes in more temperate regions, such as the English Lake District; some scientists believe they are relict populations, dating from the last ice age when conditions in northern Britain resembled those in the High Arctic today.

An integral part of the marine ecosystem is the large number of seabirds that migrate to the Arctic in summer. They include gulls, fulmars, auks, terns, divers and ducks. These birds feed on plankton and fish, of course, but since they nest on land, and provide a source of food for some terrestrial mammals, we describe them in the following section, 'Life on the tundra'.

A whale breaks the surface of the Greenland Sea between the intermittent cover of ice floes.

A pair of walrus (*Odobenus rosmarus*) grace the shores of Nordaustlandet, Svalbard. Although once largely exterminated, they are making a slow recovery, but seem to be confined to relatively few areas.

At higher positions in the food web are whales that, in order to flourish, depend on the abundance of the other component animals in the marine ecosystem. Both groups of whales, toothless and toothed species, are represented in the Arctic. The former, which are baleen whales (such as the bowhead or Greenland right whale), are adapted for feeding on zooplankton, straining these microfauna out of sea water using filters (the baleen plates). These whales typically range from 10 to 30 m in length, the largest being the blue whale which can weigh more than 100 tonnes. The largest toothed whales are sperm whales (20 m in length), followed by killer whales (10 m), and dolphins and porpoises (2–6 m). Many feed on fish and squid; killer whales (orcas), also catch large animals such as seals, by hunting in groups known as pods in the open sea. Many smaller whales, such as the beluga and the tusked narwhal, feed close to the edge of pack ice or shorefast sea ice. Large numbers can often be seen at the margins of such open water during spring.

Whales are air-breathing mammals, some of which are well-known for their intelligence, able to communicate with each other by sound and to detect objects using a type of sonar. Whales have been hunted for over a thousand years, almost to the point of extinction for a few species, largely for the oils derived from their blubber to make fuels, lubricants, soap, margarine and cosmetics. Some indigenous peoples are permitted to hunt limited numbers of whales and walrus for meat, skin and ivory. Although they are protected, some of the larger species of whale, such as the blue and sperm, are still showing few signs of recovery from over-exploitation.

Fish provide the main diet for seals, several species of which inhabit the waters surrounding the Arctic islands. Seals are warm-blooded carnivorous mammals, giving birth to their young on land or on sea ice. There are two main families of seal around the Arctic islands – walruses and true (or phocid) seals, both of which are well adapted for fast movement in the sea, but are very ungainly on land.

The walrus is solely an Arctic species and can attain a length of 4 m and weigh 1.5 tonnes. Walruses feature tusks, moustaches and bloodshot eyes. The tusks are used for breaking ice, for feeding and for providing leverage when hauling out onto sea ice. Walruses can dive to depths of 60 m, and on the sea-bed they stir up the sediment with their tusks to obtain food. They move around in small family groups, but unmated males form large herds on some Arctic beaches.

The principal species of phocid seal around the Arctic islands are the bearded, ringed, harp and hooded. Bearded seals are typically 2.5 m long and weigh over 200 kg. The attractive whiskers are probably used for sensing food on the sea bottom. Bearded seals live alone for most of the year, often close to the coast, using the sea ice as a diving platform. As the sea ice begins to break up, typically in April and May, the pups are born. Ringed seals, with their distinctive coat marked by white circles, are 1.5 m long and weigh up to 100 kg. They, too, are solitary animals, but occasionally form family groups on the ice, although they generally prefer to move away from the ice edge. Their pups are born in March or April in an ice or snow cave at the water's edge. Ringed seals have been sighted at the North Pole on several occasions. Hooded and harp seals tend to live well out to sea and rest on thicker ice floes. Hooded seals measure 3 m and weigh up to 400 kg. Harp seals are 2 m long and weigh 200 kg. The male hooded seal has an inflatable nose, used for attracting females during the breeding season. Hooded seals form a single breeding colony in the High Arctic, near the island of Jan Mayen midway between Norway and Greenland. They migrate considerable distances as the sea ice extends and retreats with the seasons. Jan Mayen also has the main concentration of harp seals

A seal surfaces in Tinayrebukta, northwest Spitsbergen. Seals are often curious of human activity, and this one swam around our vessel for some time.

The 'King of the Arctic' is the polar bear (*Ursus maritimus*). Its wide feet enable it to traverse relatively thin sea ice, as here in Hornsund, southwest Spitsbergen. (Photo: David Burkitt)

on the Arctic islands, and here pups are born in March or April. Hunting of seals for skins continues through the use of sealing vessels, under international regulation. However, there remains concern over excessive exploitation. The more sustainable approach of hunting by the Inuit yields valuable skins for clothing and rope, as well as meat and oil.

The mammal that heads the food web is the polar bear. This large carnivore commonly spends much of its life on the sea ice and adjacent waters, and so it is appropriate to regard it as part of the marine ecosystem. However, females dig dens in snow banks on land in which to give birth to their young. The polar bear competes with the Kodiak brown bear of Alaska for the status of the world's largest terrestrial carnivore. A full-grown male, standing on its hind legs, can measure 3 m nose-to-tail and weigh 500–600 kg. The polar bear's fur appears ivory, pearl, pure white, cream or yellow in colour. This variation arises from the fact that the hair is optically transparent and light is refracted. The outer guard hairs are about 15 cm long and are hard, shiny and hollow. This coat does not become matted when wet, and water can be readily shaken off before it freezes. Beneath is a thick undercoat of woolly fur, growing on black skin. Radiative energy is transmitted through the hollow outer hairs to the skin, warming the bear when out of water. A thick layer of blubber, as well as the fur, helps keep the bear warm when it is swimming, but out of water in summer this combination can lead to overheating.

A polar bear's paws, which on a male may measure about 35 cm by 25 cm,

Polar bears range widely on the sea ice off the Arctic islands. Their wide footprints are a common site on the winter sea ice which fills many Arctic fjords, such as these examples from the fjords of eastern Ellesmere Island.

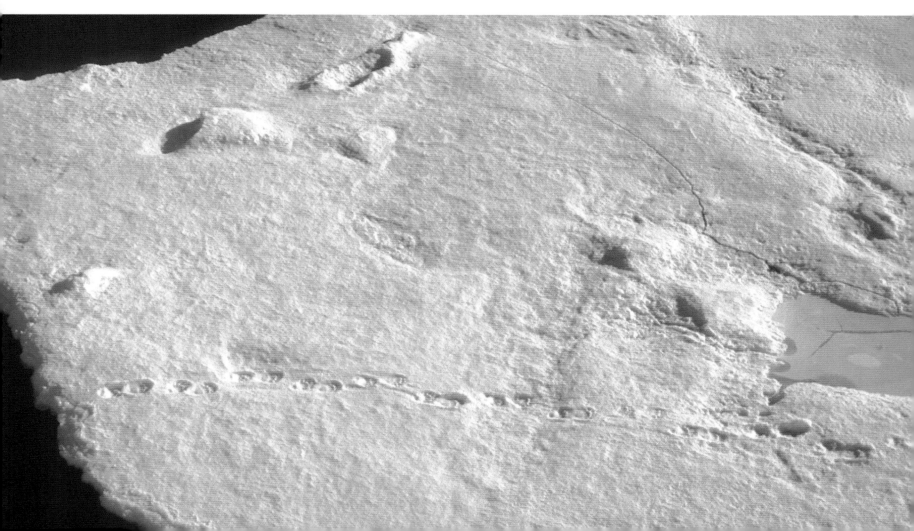

enable it to spread its weight more effectively on the sea ice. The claws enable the animal to move easily on the ice, as well as tear its prey. A bear walks at about 4 km per hour, but can reach speeds of 40 km per hour when charging a resting seal or attacking a human. Bears are extremely agile and dextrous. They are deft enough to snatch a lemming from tundra vegetation, but can also knock a beluga whale senseless with a foreleg, or toss a 200 kg seal into the air. Polar bears feed mainly on ringed seals which they catch by stalking until one pops its head through a breathing hole in the ice, and is stunned by a swipe from the bear's paw, followed by a smashing blow with its snout. A bear may display considerable patience, waiting several hours at a seal-hole, lying on its chest downwind of the hole, out of the seal's line of sight. Bears prey on other species apart from seals, including beluga and narwhal; they have also been known to kill walruses, musk oxen, hares and geese. When food is scarce, bears will eat anything, including people, or carrion.

Bears are solitary animals and only get together to mate, in April or May. A male bear will sometimes kill a mother and cubs if they subsequently encounter one another. Female bears occupy snow caves in winter, especially when food is scarce, to conserve energy. They prefer sites where deep snow drifts form in early autumn. They will excavate a hole big enough in which to turn around, reached by

a 3 m-long tunnel. The female here gives birth to a couple of cubs within the den in January or February, and the family only emerges in March or April. Until then, the mother lives off her body fat, and the cubs rely on her creamy milk. The family remains together for about two years.

Polar bears are noted for wandering considerable distances over the sea ice, far from land. Bears have even been spotted within about 100 km of the North Pole. Yet bears maintain discrete populations, returning to their regular hunting and denning areas annually, such as Kong Karls Land in Svalbard, or the eastern hills of Wrangel Island. The Svalbard population wanders across the sea ice as far as the east coast of Greenland, and to Franz Josef Land to the east. One of these bears, tracked by radio after being tagged, was found off East Greenland one year later, a straight-line distance of 2,000 km from its Svalbard home. Another group of bears roams through the Canadian Arctic Archipelago. Yet another has its range in the southern part of Hudson Bay, far south of the Arctic Circle. As the bear wanders across the sea ice, it is often accompanied by the Arctic fox, which feeds on the remnants of a killed seal, along with a range of scavenging birds. A fox has been sighted as close to the North Pole as 140 km, even though it is normally a land animal.

The polar bear was hunted to the verge of extinction in parts of its range, but in 1973 an international agreement for the protection of the animal was signed by Canada, Denmark, Norway and the USA, with the then USSR joining the following year. Only a strictly limited quota of bears may be taken locally, while a small number may be killed for *bona fide* scientific purposes. Most people are allowed to shoot bears only if attacked. Protection has proved highly successful, and in some areas, such as Svalbard, numbers are thought to have doubled since 1973. Bears to some extent have now lost their fear of humans and, with the increasing number of scientists and tourists visiting the Arctic, they have become a hazard that must be taken seriously. Official posters warn visitors of the behavioural characteristics of bears and explain what to do if one attacks. The text of one such poster, produced by the Norwegian authorities for visitors to Svalbard, is reproduced in the accompanying box. Visitors are required to carry a weapon at all times outside the main settlements. Recent years have seen a number of fatalities from bear attacks, all attributable to visitors having had insufficient protection.

Ministry of the Environment/Norwegian Polar Research Institute

It attacks without warning –

KEEP YOUR DISTANCE!

Polar bears may be very dangerous, especially when hungry or feeling threatened. But they are also inquisitive animals, wanting to investigate unknown items or situations, but not necessarily with aggressive intentions. Then, at close quarters, they may suddenly feel threatened and attack.

Watch for these signals

A polar bear's behaviour may tell us if it is merely inquisitive or intends to attack. A curious bear will often approach slowly, stopping now and then, craning its neck, sniffing. It will then move its head from side to side, up and down.

An aggressive polar bear will show more determination, and may often attack without warning. Sometimes, however, it may give an aggressive signal in advance, for example by blowing violently through its nose like an angry bull, or by gnashing its teeth with a smacking sound. Then it's time to be on guard! A polar bear will attack very quickly. At a rapid trot or in great, supple leaps it will aim directly at its prey. If possible, fire warning shots or throw flares on the ground around the bear.

The following precautions should be taken

1.   Always carry a weapon when travelling in polar bear areas.
2.   Do not attract bears by putting out food. Never feed them from a ship or from your cabin window.
3.   In the field, place your garbage at least 100 m away from the camp, directly in view of your tent opening or cabin door. This may enable you to see a visiting bear in time.
4.   If you feel you need further information, contact the Governor of Svalbard (Sysselmannen).

Safety rules

Always carry a weapon when travelling in polar bear areas. Keep calm if you see a bear approaching. Keep your weapon ready and withdraw slowly if the bear comes closer. If you find yourself in a dangerous situation, throw down your hat, mitten, scarf, etc. The bear will usually stop to sniff at the garment, which will give you time to reach safety.

The polar bear is totally protected, and should only be killed in self-defence. If you must shoot, do not aim at the head – it is easy to miss. Aim at the chest. Even if the first shot is not fatal, the bear will stop for a moment, giving you enough time to fire again. Report to the authorities (Sysselmannen) if you have had to kill a bear. Find out what sex it is, and take care of the skull and skin.

Life on the tundra

The key to understanding the life cycle of terrestrial plants and animals in the Arctic is an appreciation of the ecology of the tundra. The term tundra is a Lapp word referring to the treeless plains of northern Europe, but is more broadly applied to the environment that exists north of, or above, the tree-line. Thus, all the Arctic islands lie within the tundra zone. We approach this topic, as for the marine ecosystem, by considering the interactions between the different components and the organisation of the food web. Starting with the development and nature of soils, we explain how progressively higher plants become established, shelter insects and small mammals, and provide a source of food for birds and grazing mammals. We also examine the connection with the marine realm, and consider how sea birds contribute to the welfare of land animals.

The presence of permafrost (Chapter 6) is the main factor that controls the development of soils. Permafrost is rock-hard. Thus, the roots of plants are unable to penetrate far into it. Liquid water is only available to plants in summer, when the so-called active layer thaws out to a depth of 1–2 m. Soils in permafrost regions are largely the product of frost-weathering of bedrock or unconsolidated sediments laid down by rivers, glaciers or as beach deposits. These mineral soils are first colonised by bacteria and algae. Higher plants then take advantage of the nitrogen that becomes fixed in the soil by certain bacteria. However, the decomposition of twigs and leaves is very slow, and little humus is produced. The soils are thus thin, acidic, commonly waterlogged and poorly aerated, except where they receive added nutrients from the droppings of nesting birds. The vegetation traps water and moderates the microclimate, encouraging tiny insects and nematode worms to establish themselves. These animals, in turn, keep minerals in circulation, and thereby improve the soil's productivity. Soil development is exceedingly slow. For example, areas exposed for a century following glacier recession in Svalbard are still dominated by organic-poor soils with only isolated flowering plants and few insects. Even areas exposed since the last glacial maximum, around 18,000 years ago, commonly have only patchy soils and discontinuous plant cover. This is because soils are subject to extremes: if too dry, wind removes the finer particles; if too wet, the freeze–thaw process constantly disturbs the soil, creating the typical patterned ground (Chapter 6). In the drier parts of the High Arctic, such as North Greenland and Ellesmere Island, the absence of moisture limits soil formation and vegetation

Bright green moss, lichen and flowering plants characterise scree slopes beneath cliffs which are sea-bird nesting sites in many coastal areas of the Arctic. In the photograph, from Krossfjorden in northwest Spitsbergen, the bright green of this highly nutrient-rich location stands out from the surrounding mountainsides.

Lichens, East Greenland.

growth. Such conditions are termed a polar desert, an environment dominated by wind erosion and deposition, and frost action.

Despite the poor quality of tundra soils, a wide variety of plant species is to be found on the Arctic islands, including algae, lichens, mosses, sedges, rushes, grasses, herbs, shrubs and dwarf trees. Sheltered valleys, and especially gentle south-facing slopes, where summer temperatures may reach 15–20 °C, are favourable areas for vegetation, which can grow dense enough to seriously hinder travel on foot.

Tundra plants generate their own microclimates, and ground temperatures may be as much as 10 °C higher than the air temperature. Small invertebrates benefit from these warmer conditions. Tundra plant communities vary considerably, and several classifications relating to latitudinal and altitudinal zones have been devised, reflecting the diversity of landscape and climate. However, the overriding factor controlling where plants occur is the amount of summer warmth

Yellow mountain saxifrage (*Saxifraga aizoides*) is one of the latest species to flower in the High Arctic summer.

Soft green cushions of moss campion (*Silene acaulis*) grace the drier parts of the tundra landscape.

Fragile fern (*Cystopteris fragilis*).

Svalbard poppy (*Papaver dahlianum*).

Reindeer moss (*Cladonia rangiferina*).

available. Each plant species has its own specific requirements and tolerances. If these change, then plant reproductive success and even survival are compromised.

Extensive areas of the High Arctic are characterised by rock deserts and barrens. In rock desert (or fell-field) habitats only scattered tufts and cushions of plants are present, and continuous ground cover is absent. The soils are poor and dry out quickly. Lichens form patchy cover on rocks, but there is also an attractive array of flowering plants such as Arctic poppy, moss campion, whitlow-grass and saxifrages, as well as grasses, sedges and ferns. Species such as purple saxifrage and Arctic poppy are among the first to colonise areas from which glaciers have receded, and do so within a few years. Barrens consist of smoother areas of fine-grained rock particles, but although the species diversity is low, plants are able to form more continuous colonies, frequently arranged in the form of islands or strips, and reflect the nature of the patterned ground that supports them. Typical species are purple saxifrage, Arctic poppy and mountain avens.

Tundra grasslands are found in the southern parts of Greenland or sheltered inland areas of Svalbard, the New Siberian Islands, Wrangel Island and in parts of the Canadian Arctic islands. Here, protected, sunny slopes support a variety of grasses, mosses and lichens. Such slopes provide sustenance to musk oxen (in Canada and Greenland), reindeer (in Svalbard) and other herbivores, and also provide nesting sites for birds such as waders. Slopes that carry a protective snow cover, have good soil, and are well-drained as the snow thaws, are referred to as herbslopes. They support a rich flora of flowering herbs, including saxifrages, moss campion, Arctic poppy, and the occasional shrub-like birch or willow. In many areas, snow remains for the entire year or may be absent for only a few weeks. Such conditions give rise to a distinctive snow-patch community, comprising mosses and

Purple saxifrage (*Saxifraga oppositifolia*) is one of the first flowering plants to emerge from the snow in early summer.

Arctic willow (*Salix glauca*) in flower.

Broad-leaved willow herb (*Chamaenerion latifolium*).

low shrubs such as Arctic willow. Mosses are efficient absorbers of solar radiation, and generate a warm microclimate in which insects can thrive. Green and red algae may accumulate in old snow patches, and colour the snow patches in some areas.

The presence of permanently frozen soil inhibits drainage on flat ground, so many Arctic areas are typified by wetlands. Marsh plants and mosses act like sponges in such terrain, whilst willow, birch and other shrubs help to stabilise the banks of watercourses. Rotting vegetation produces peat-bogs after a few hundred years. Aquatic plants thrive along lake shores, despite the annual freeze-up for all but two or three months a year. In fact, near sea level, there are few places where plants cannot grow. Even bare rock surfaces support vegetation, whether it be a covering of green, orange or black lichen, or flowering plants occupying crevices.

The vegetation supports insects, mites and tiny spiders. Springtails, living on rotting vegetation, are especially numerous, and provide food for spiders, beetles and predator mites. Butterflies and moths occupy sheltered vegetation. Boggy areas provide the breeding ground of that curse of the Arctic, the mosquito. Although not

Snow melting in early summer gives rise to marshy ground, dominated by mosses and small shrubs. This wetland area is on Ymer Ø in East Greenland.

Hairy lousewort (*Pedicularis hirsuta*).

Arctic wintergreen (*Pyrola* sp.).

Nodding lychnis (*Silene apetalum*).

Arctic hare (*Lepus arcticus*) on the tundra of Ymer Ø, East Greeenland.

usually a nuisance to humans in the far north, clouds of mosquitoes emerge from larvae in lakes of the sheltered valleys of Baffin Island, Greenland and Svalbard. Female mosquitoes seek the blood of mammals to obtain protein for the production of their eggs, and are noted for their irritating bites and buzzing sound.

Tundra vegetation provides cover for the lemming, one of the few rodents to be found in the Arctic islands, and these inhabit some of the Canadian islands, parts of Greenland and the Siberian islands, but not Svalbard. Adult lemmings are typically 10–15 cm long, excluding the short stubby tail, and have a dense fur. Their numbers fluctuate dramatically from year to year, depending on the weather conditions and the quality of the vegetation that serves as their food supply. Superimposed upon this is a regular cycle of population growth and collapse: typically this cycle is four years, but varies from region to region. The availability of lemmings has a profound effect on the breeding success of species that feed on them, such as foxes, jaegers and snowy owls. Lemmings live underground, and remain active in tunnels under the snow pack throughout the winter, feeding on the buried vegetation. Breeding begins early in life, and in years of abundant food, females produce the first of four or five litters in March.

The tundra of Greenland and the Queen Elizabeth Islands is home for a large species of hare, the Arctic hare, which remains white throughout the year. Typically, adult hares weigh 5 kg, and form loose family groups of a handful of members to as many as 100. They roam over the tundra throughout the year,

Acquaintance with a musk ox (*Ovibos moschatus*) at close quarters is an unnerving experience. This animal, on Ymer Ø in East Greenland, charged the photographer without warning, but fortunately stopped a few metres away.

propelling themselves by a series of hops interspersed by bouts of running. They manage to live throughout the winter on sparse vegetation exposed by wind, whilst their white camouflage against the snow provides concealment against predators. Breeding takes place in spring, each female producing a litter of some half-dozen leverets.

The tundra vegetation also supports large grazing animals: musk ox, caribou and reindeer. The musk ox is the largest herbivore found on the Arctic islands, and is perhaps the species best-adapted of all to survive on the tundra. The musk ox is one of the few large animals to have survived the ice ages in the North American Arctic. Many of its ice age contemporaries, notably the woolly mammoth, have become extinct. Today, musk oxen are found only in North and East Greenland, the Canadian Arctic Archipelago, and on Wrangel Island, where the species was introduced in 1975. Attempts to introduce them to Svalbard in the 1930s failed after half a century of struggle, although numbers peaked at 50. The musk ox has downward-curving horns, and stands about 1.5 m at the shoulder, weighing up to 300 kg. Along with the horns, its most striking characteristic is its long, flowing skirt of soft 'guard' hair, especially when it is moulting in large clumps and streamers in summer. Beneath this outer hair is an undercoat of fine woolly hair that, weight-for-weight, is far warmer than the wool of sheep. Musk oxen move around in small groups of two or three, but sometimes in herds of up to thirty. They are entirely at home on the tundra, where they favour relatively lush moist meadows, grazing on flowering plants, grasses, sedges and mosses.

Musk oxen on the move are an impressive sight. They can gallop for several kilometres, often shoulder-to-shoulder, turning or stopping as though a single unit. Although musk oxen remain passive towards each other for most of the year, during the rutting season in late summer, competing males will often charge, locking horns or head-butting each other with tremendous force. The loser may occasionally sustain fatal injuries. Subsequent mating takes place in late summer, with calves being born eight months later in the spring. The calves both suckle on their mother's milk and eat vegetation from an early age.

When threatened, musk oxen arrange themselves into a defensive line or circle. Such a posture, with an array of horns pointing outwards, provides a near-impenetrable defence against their major predator, the wolf. Only if the wolf pack can cut off one of the younger or weaker members of the group are musk oxen vulnerable. However, in the face of rifles, the same defensive arrangement has accounted for their downfall at the hands of humans, and herds throughout the Arctic were greatly reduced or eliminated. Early European explorers in particular over-exploited the musk ox. For example, they were wiped out from areas such as Banks and Victoria islands. Even as late as the 1950s there were only a few individuals, but in the following thirty years the Banks Island population recovered to around 18,000 animals. Subsequently, there has been renewed pressure from the Inuit in certain areas, and musk oxen have been killed because they competed with caribou for the best grazing. In the past, the musk ox has also proved to be a valuable resource for the Inuit: ladles have been made from the horns and tools from the bones, whilst the hides have been used as insulating rugs. In recent decades, the musk ox has been largely left unhunted, and has begun the slow recovery to a sustainable population. Musk oxen need to be watched carefully as solitary bulls, pushed out of the herd, have been known to attack humans, as indeed one of the authors has experienced.

The caribou of the Canadian Arctic islands, and the reindeer of Eurasian islands and Greenland are actually the same species, although there are many local subspecies. Caribou are the same height as musk oxen and weigh approximately 250 kg when fully grown, whereas reindeer on Svalbard are only a metre high and weigh less than half the weight of caribou (although size is larger in the Norwegian and Russian mainland varieties). Both males and females carry antlers that are shed annually in the autumn. Caribou and reindeer of the Arctic islands, in contrast to their mainland counterparts, only form small herds, typically around a dozen members, and are often solitary. They graze throughout the year, eking out an existence from lichens and mosses where the snow cover is thin, but taking full advantage of the richer vegetation of summer. Calves are born in the spring, ready to take advantage of early summer bounty. Like many Arctic species, these deer were subject to slaughter on a large scale, especially during the nineteenth century when European hunting was at its peak. Populations in certain areas have been recovering well since controls on hunting were introduced, although adverse weather conditions, such as the ice storm of 1977 on Banks Island, which prevented many caribou from reaching food, can result in population crashes.

Other animals benefit from the activities of musk oxen, caribou and reindeer. Snow buntings line their nests with wool and hair. Although nesting birds such as jaegers and skuas become irritated when their nests are threatened by trampling hooves, the pawing of the ground through snow and ice in winter allows hares and ptarmigan access to food supplies. The larger animals also stir up ground in summer, allowing birds to feed on the uncovered insects. On death, the carcases of

The 'pot-bellied' nature of this reindeer (*Rangifer tarandus platyrhynchus*), observed near Kapp Linné in western Spitsbergen, indicates that the summer food supply has been more than adequate.

large animals provide sustenance for scavenging mammals, birds and insects.

The tundra maintains sparse populations of several species of wintering bird. The most striking of these is the snowy owl, which, as its name suggests, is all white. Snowy owls live in northern Greenland, the Canadian Arctic islands, Novaya Zemlya and Wrangel Island, but are not generally seen in Svalbard. They feed on lemmings, ptarmigan and other small birds. Their numbers fluctuate dramatically depending on the abundance of the small rodents. Other raptors include gyr falcons and peregrine falcons that breed along the coasts of the Canadian Arctic Archipelago and Greenland. Ravens are known to breed in Greenland, the Canadian

One of the few year-round resident birds on the Arctic islands is the ptarmigan (*Lagopus mutus hyperboreus*). After the autumn moult its feathers change from brown to white, providing good camouflage irrespective of the conditions. This is the Svalbard subspecies, observed near Ny-Ålesund in northwest Spitsbergen.

Arctic Archipelago and Wrangel Island. Their diet consists of anything from berries to meat, often 'tidying up' after an animal kill. The rock ptarmigan lives as far north as land reaches, its colour changing from speckled brown in summer to white in winter. The relatively thin snow-cover enables it to eke out a living through the winter months. In the severest weather it burrows into snowdrifts for insulation and obtains seeds, shoots, leaves and berries. In summer, ptarmigan nest on clumps of sparse tundra vegetation, and it is easy almost to step on an adult and its half dozen or more chicks, as they are so well camouflaged and do not readily move away from intruding humans. They are easy prey for foxes, raptors and other

Black and white male common eider (*Somateria mollissima*) competing for the attention of brown females in early summer in Spitsbergen.

hunters. The smallest wintering species is the snow bunting, a white sparrow-like bird with black markings, which lives and breeds on all the Arctic islands, as well as in the mountains of the temperate zone, such as Scotland and Norway. Its main food is grass seed, but in summer it benefits from the supply of insects.

The tundra plains and cliffs are transformed by the arrival of summer visitors that include both aquatic and terrestrial birds. The largest land-based birds belong to the goose family and include the barnacle, pink-footed and Brent species, which arrive in large flocks to nest on tundra plains or even on cliffs. The migratory movements of barnacle geese are particularly well understood, since they have been monitored using small transmitters fitted on their backs. For example, one population that breeds in Kongsfjorden, northwest Spitsbergen is known to migrate to and from the Solway Firth on the Scottish/English border. Ducks are represented especially by the common eider. This species is famed for its soft, superbly insulating, feathers or 'eider down', which it uses to line its nest. The common eider breeds on beaches, around lagoons, or on low coastal cliffs on most of the Arctic islands. As soon as the chicks are hatched, they are taken to water for their first swim. Eiders in coastal waters form small groups of brown females and striking black-and-white males. The cooing sound emitted by the drakes is one of the most evocative sounds one can hear in the Arctic. Other land-based birds include a variety of waders (turnstone, sandpipers, stint, sanderling, knot and dunlin), all of which breed on the Arctic islands.

In terms of sheer numbers, it is the visiting sea birds that render the Arctic islands an ornithologist's paradise. They feed on marine plankton and fish, but are beneficial to the terrestrial environment by adding nutrients to the soil, and providing a source of food for predators. Gulls are particularly abundant and nest in large colonies on cliffs. They include the herring gull and kittiwake, which have spread north to places like Svalbard as the climate has become warmer over the last hundred years. The huge glaucous gull, with its 1.5 m wingspan, is a predatory bird, feeding on other birds and their eggs, as well as scavenging after mammals, following a kill. This gull can often swallow an eider duckling or tern chick in one gulp.

A common summer visitor to Svalbard is the barnacle goose (*Branta leucopsis*). They nest on bare open ground, close to the coast, as this view at Ny-Ålesund, northwest Spitsbergen, shows. The geese in this area originate from the Solway Firth in western Britain. Their young are vulnerable to predation by foxes.

All these species nest on coastal cliffs where they are safe from predatory foxes. In contrast, ivory gulls often choose to nest on high mountain cliffs inland. This beautiful, often silent, bird is pure white, and forms a welcome companion, as it accompanies sledging parties as they cross inland icefields.

The arctic tern is a small elegant seabird, adapted for fast, manoeuverable flight. However, it can be vicious, often attacking humans and other animals with little provocation, especially near its nests. It seems to have a perverse desire to nest within human settlements, and then attacks all-comers as they pass by the nest, aiming for the head and not infrequently drawing blood. Also aggressive are Arctic skuas and long-tailed jaegers which nest on flat tundra and stony deltas during summer, but live on the ocean during winter. They commonly attack humans from behind, often in pairs, but then feign injury to draw the perceived threat away from the nest. Skuas also attack other birds, forcing then to disgorge food.

The widely distributed fulmar is a species of petrel with a wing-span of a metre or more, and a remarkable life-span of about 70 years. Fulmars breed extensively on sea cliffs in Svalbard, Greenland, the Canadian Arctic Archipelago and the islands off northern Russia. Some nesting colonies are established in mid-winter, but most birds arrive in spring when the winter sea ice breaks up. A single egg is laid in June and the chicks leave the nests in September. Fulmars feed on small fishes and crustaceans, but also scavenge for offal from fishing vessels. The fulmar is a superb flier, characteristically skimming over the sea or closely following the line of the sea cliffs. Phalaropes are wading birds that nest on the tundra areas of Baffin Island, Greenland and several other Arctic islands. They lay their eggs in May and, unusually among birds, the male incubates and rears the chicks. Phalaropes feed on insects and freshwater plankton in streams and lakes. After breeding they spend the winter far south in the Atlantic and Pacific Oceans.

Auks (or alcids) are small black and white diving birds, represented in the Arctic islands by the little auk, guillemot and puffin. They gather their food at sea by diving and swimming under water. Little auks or dovekies nest in vast numbers on coastal and inland cliffs. Here, they create an incessant chattering sound and fly around the cliffs in swarms, wings whirring at high speed. The locations of these colonies may easily be detected by smell, even through thick fog. Guillemots are more slender black and white birds which nest on sea cliffs, but also spend a lot of time on ice floes and small bergs. They can dive to well below 100 m in search of fish. Puffins have striking red and yellow bills, although their plumage is black and white. They nest both on cliffs and in burrows, but are less common than other

One of the most elegant birds in the Arctic, yet at the same time the most aggressive, is the Arctic tern (*Sterna paradisaea*). Nesting in Spitsbergen in summer it migrates to the Antarctic after breeding to take advantage of the summer in the southern hemisphere too.

Nesting on tundra away from the coast is the long-tailed jaeger (related to the Arctic skua). Here on Axel Heiberg Island in the Canadian Arctic archipelago, a jaeger is attacking the photographer who had inadvertently approached too close to its near-invisible nest.

alcids in the Arctic islands. Foxes are commonly seen below the cliffs of nesting birds, where they can obtain a good supply of chicks that have fallen from the nests.

Several species of diver, notably the great northern diver (known in North America as the common loon) and the red-throated diver, may be observed in the Arctic islands. They are solitary water-birds, capable of remaining underwater for up to three minutes, and diving to depths of at least 60 m. Great northern divers breed on lake shorelines in Greenland, Jan Mayen and Bjørnøya. The red-throated diver also breeds far to the north – in Greenland, Ellesmere Island, Svalbard and the Russian Arctic islands.

The largest carnivore of the Arctic islands is the polar bear. As a largely marine-

Black guillemots resting on an ice floe in a Spitsbergen fjord.

based animal, this has already been described above. However, it is also known to venture inland where it might attack the occasional musk ox, reindeer or human. Incursions inland most likely are a means of switching from one coast to another, as from the west coast to the east coast of Spitsbergen when the sea ice limit recedes northwards and eastwards, and it is no longer able to hunt effectively in the west. In fact a bear has been observed near the top of Newtontoppen, Svalbard's highest peak (over 1700 m). Another has been observed on the Greenland Ice Sheet at over 2000 m above sea level, and 50 km inland.

The grey wolf is an elusive, solely meat-eating animal found in the Canadian Arctic islands and Greenland, although absent from Svalbard and the Russian Arctic islands except southernmost Novaya Zemlya. Its large size (50 kg) and ability to hunt in pack make this a formidable foe to species such as caribou, although it generally is selective in weeding out sick or weak animals. On the Arctic islands pack-size is smaller than on the mainland, rarely comprising more than half a dozen adults. Wolves commonly move around in pairs or singly in the far North. More than most animals, the wolf has been subject to extreme persecution. Even today, it is often shot on sight, except where totally protected in the national parks. Wolves give birth to litters of five to ten pups in May, usually under cover of rock

A frequently-sighted year-round resident of the Arctic islands is the Arctic fox (*Alopex lagopus*), prized for its dense white winter coat. This fox was photographed close to the Canadian weather station at Eureka on Ellesmere Island.

crevices or beneath boulders on the tundra. The pups remain with the family group for over a year, learning the skills necessary for independent survival. In addition to feeding on sick larger animals, wolves feed on small mammals and birds, and eggs. Wolves do particularly well when lemmings are abundant.

The Arctic fox is a much smaller member of the dog family, rarely weighing more than 8 kg. Its range is completely circumpolar. In summer the fox appears lean and slender with a short grey-brown and fawn coat and bushy tail, whereas in winter it carries a voluminous white coat, much prized by trappers. Some populations have a bluish grey coat in winter, notably in Greenland. Arctic foxes give birth to six to eight cubs in May, though many may not survive. However, in years of abundant food, a second litter may be born in late summer. The fox, unlike the wolf, is a solitary animal, feeding on small mammals such as lemmings and breeding birds. Even large birds like barnacle geese are fair game. Foxes are often attacked by terns and skuas on the tundra, but they seem to take this in their stride. Foxes can often be seen loitering beneath bird-nesting cliffs, waiting for young gulls or auks to fall to the ground. They store food in crevices, out of reach of other predators, as insurance against poor food supplies in winter. They are also known to follow polar bears across the sea ice in winter for scraps.

Survival of arctic fauna and flora, and future prospects

Compared with other parts of the world, Arctic ecosystems are relatively young. In the tropics, ecosystems have evolved over millions of years but, apart from a few 'refugia', most Arctic ecosystems have evolved only over ten to twenty thousand years, following the retreat of the last great ice sheets. In some areas, particularly those covered by glaciers until recently, colonisation by plants and animals has barely begun. Colonisation is a slow process. Plants have to adapt to a short growing season with 24-hour daylight in summer and 24-hour darkness in winter. Animals have had to evolve breeding and feeding strategies that depend on uninterrupted light. Life has to contend with unpredictable and violent weather. A June snowstorm or an early August freeze can destroy an entire generation of young birds, seals and caribou, while the carnivores which depend upon them suffer too.

Animal population crashes are common. For example, on Wrangel Island the population of snow geese declined over the ten-year period to 1975 from 400,000 to fewer than 50,000 as a result of a succession of spring snow storms, but had recovered to 300,000 by 1982. Again, in the autumn of 1973, an October rainstorm in the Canadian Arctic Archipelago created a layer of ground ice that musk oxen could not break through to feed, resulting in the starvation of 75% of the population. Once again, however, recovery from this disaster was dramatic. Thus, biologists view Arctic ecosystems as 'stressed' or 'accident prone'. At the same time, animal species are remarkably resilient, and recover well from natural disasters.

Of course, animal populations have suffered greatly at the hand of humans, although rarely, it has to be said, at the hands of the indigenous peoples. Fortunately, partial or complete protection nowadays prevents the carnage that was typical of the past. Species like the polar bear, musk ox, reindeer, fox and walrus have recovered well in recent decades. In contrast, there are few signs of recovery of the large whales, such as the sperm and blue, which were subject to over-exploitation in the eighteenth and nineteenth centuries.

Despite these success stories in conservation, there remains the more insidious threat from industrial pollution. Toxic chemicals, originating in the industrial zones of North America and Europe and transported in the atmosphere, have been found in polar bear livers and the shells of arctic tern eggs. These chemicals could affect the fertility of animals and induce deformities in the young. The nuclear power station disaster at Chernobyl, in the Ukraine, resulted in excessive levels of radiation in Svalbard reindeer. Damage to the atmosphere is evidenced by the formation of an incipient ozone hole over the Arctic in a similar way to the serious and well-known case over the Antarctic – the result of halocarbons entering the atmosphere from refrigerants and propellants in spray cans. Increased ultraviolet radiation resulting from damage to the ozone layer is likely to damage shallow marine and terrestrial ecosystems. Fortunately, an international protocol is stabilising such emissions, and it is expected that the atmosphere will recover slowly.

Perhaps the greatest threat of all is global warming, resulting largely from the emission of carbon dioxide through the burning of fossil fuels. Already CO_2 concentrations exceed that of anything recorded by the atmosphere in the last 2 million years, and unprecedented temperature rises of up to about 4–5 °C are predicted for the Arctic by climate modellers in the next hundred years. Such changes will accelerate the reduction of sea ice and glaciers and begin thawing the tundra, impacting on the plants and animal communities of the Arctic islands and in the waters surrounding them.

Chapter 9

Indigenous peoples, exploration and environmental impacts

The Arctic is often perceived as a remote wilderness, and, while this is true of very large tracts of territory, human impact has also been considerable. Several Arctic islands have been the home for native peoples for thousands of years, but they have tended to live in harmony with nature and have left little evidence of their presence. Exploration and exploitation came with expeditions from the South, notably in the last three centuries, and the flora and fauna of some areas was damaged. During the twentieth century the Arctic became a place of confrontation between the great powers in the Cold War. In the last twenty years, tourism has expanded greatly, and with it a widespread awareness of the fragility of the Arctic ecosystem – the complex set of interactions between plants, animals and the soil. The new Millennium begins on a more optimistic note – recognition of the need for sustainable development, alongside an efficient infrastructure to allow visits by tourists, researchers and educationists. But behind all this is the more insidious threat of global warming, the consequences of which are only now beginning to be understood.

Indigenous peoples

The famous old Norwegian coastal vessel, *Nordstjernen*, which has been plying the waters of Svalbard for several decades, is dwarfed by the mountains bordering Kongsfjorden in northwest Spitsbergen.

The indigenous people of the Arctic islands are the Inuit (in Greenlandic, Inughuit), formerly known as Eskimo or Esquimaux. Today, Inuit inhabit the islands of the Canadian Arctic and Greenland. The earliest evidence of their forbearers is found about 4,500 years ago, and artefacts of the widespread Dorset culture, dating from about 3,000 years ago, have been identified at a number of sites in the Canadian Arctic. Archaeologists have found evidence that the ancestors of present Inuit first entered Canada across the sea ice from what is now north-eastern Siberia, across the Bering Strait, and continued to Greenland *via* Nares Strait. The traditional Inuit culture is one of co-existence and interaction with nature, with hunting of Arctic sea and land animals providing not only food, but also the raw materials for heat, shelter and tools. Heat was obtained from seal and whale blubber, often burned in lamps fashioned from local soapstone. Tools were made from bone. Shelter, and of course clothing, was from the skins of hunted animals. In addition to the igloo, constructed from cut blocks of snow, the Inuit also made shelters with rock or sod foundations and hide stretched over the huge ribs of whales. Both could be constructed relatively rapidly, and were well suited to the semi-nomadic life of the hunter.

The Inuit of northern Canada and Greenland often hunted at the ice edge – that is, where open water is present at the edge of sea ice. In these areas, both polar bears and large sea mammals congregate: seals, and whales ranging from the large bowhead to the smaller tusked narwhal and beluga. The Inuit, through their need to travel and hunt on sea ice, acquired great expertise in judging the condition of the ice and whether it was thick enough to bear weight. In fact, the Inuit language, Inuktitut, contains more words describing ice than any other.

While some Inuit still set up isolated spring and summer camps, most now live in established settlements. For example, some families who winter in the East Greenland settlement of Scoresby Sund, establish summer camps in Kangerlussuaq Fjord, some 200 km to the south. The preservation and indeed the exercise of traditional Inuit skills in hunting and travelling therefore continue. Recognising the need to support their life-style, the international community still allows Inuit to hunt otherwise protected species such as the polar bear and musk ox. Indeed, the establishment of Nunavut Territory from part of the former Northwest Territories of Canada in 1999, together with elected representatives for self-government, is important recognition of the status of the indigenous population. Nunavut encompasses the bulk of the Canadian Arctic islands, an area of about 2 million km^2, with its capital and administrative centre in Ikaluit at the head of Frobisher Bay in southern Baffin Island. Nunavut has a population of only about 27,000. In Greenland too, self-government by the native people has been established.

Inuit children drag a dead seal across the snow, followed by husky puppies, Resolute Bay. The Inuit use seal not only for food, but also for durable clothing. Additionally, seal blubber can be used for cooking (Photo: Fritz Müller Collection).

Inuit with their husky dog-teams mingle with military personnel near the US Airforce base of Thule in Northwest Greenland (Photo: Fritz Müller Collection).

Early exploration and exploitation of living resources

The first European explorers of Greenland were the Norse. Erik the Red founded a colony in Southwest Greenland in AD 985 which grew to about 2,000 people. Indeed, it was Erik who chose the name Grønland (*Green*land) to encourage his compatriots to settle with him on this largely ice-covered island. However, the way of life based on the grazing of hardy domestic animals from Europe, rather than indigenous hunting techniques, foundered under deteriorating climatic conditions, perhaps exacerbated by the plague. By the early fifteenth century the community had become isolated from Europe and failed.

Elsewhere in the Arctic, indigenous peoples such as the Sami and Komi of

Lapland did not settle the archipelagos north of Eurasia. Instead, among the first to explore the islands around the Barents Sea, particularly Svalbard and the western coast of Novaya Zemlya, were probably Norse. From the sixteenth century onwards, whalers and sealers went north, especially from England and Holland, in search of wealth from the Arctic seas. The whalers and sealers were often secretive with their knowledge of northern waters, because it gave them a competitive advantage over the less experienced. A number of fjords in Spitsbergen became sites for occasional overwintering, and for the treating of whale carcasses. A relatively large community of whalers grew up on Amsterdamøya, an island in northwest Spitsbergen, in the seventeenth century. Less well known were the Russian Pomors from around the White Sea. Graves on Novaya Zemlya are the main sign of their visits. Many individual hunters and trappers also operated in Svalbard in the nineteenth and early twentieth centuries, and indeed several trappers still live in isolated locations on Spitsbergen today. The fur and skins of polar bear and Arctic fox were their principal sources of income. Operating in such harsh environments was only for the experienced. The remains of a jammed rifle in the Svalbard Museum in Longyearbyen are the only reminder of one early trapper who disappeared. Huts, including those of former trappers, are scattered around the coasts of Svalbard and Greenland as well as several other islands. They now serve as convenient refuges for modern travellers, although some suffer from recurrent damage by polar bears.

The search for the Northeast and Northwest passages

The more formal early exploration and mapping of the Arctic islands took place for a number of reasons. Many were interested in the search for a northern passage for shipping to the Far East, especially for trade with China, via either the Northwest Passage through the Canadian Arctic Archipelago, or the Northeast Passage north of Siberia. Such routes would be much shorter than sailing around Cape Horn or the Cape of Good Hope.

The leading sea-faring nations of the sixteenth century, England and Holland, initiated the search for these alternative routes. Attempts to find a Northeast Passage began with Hugh Willoughby and Richard Chancellor in 1553, but they made little progress. In 1596, the Dutch navigator, Willem Barents discovered and named Spitsbergen after the many sharp peaks he saw on the island's west coast, and demonstrated the commercial potential of whaling. Further efforts to find a passage failed until, as late as 1879, the Swede Adolf Erik Nordenskiöld succeeded in

The grave of a sixteenth century whaler in Krossfjorden, northwest Spitsbergen. This was a sheltered, if bleak anchorage used for overwintering by English whaling ships in particular. The wooden coffin was once buried, but has come to the surface through the process of frost heaving, as the thin soil freezes in winter and melts each summer.

A number of small huts are found around the coast of Spitsbergen. Some were the homes of former trappers, while others are maintained as shelters for short-term visits or emergencies. This one lies near the shore of Krossfjorden at the entrance to Tinayrebukta, northwest Spitsbergen. With a small stove burning driftwood, such huts can be very cosy places in bad weather. Outside, driftwood has made a useful table and chairs for use on the few relatively warm days of summer.

sailing eastward along the Siberian coast to the Bering Strait. The Northeast Passage (now known as the Northern Sea Route) is used today as a transport route by Russian ships, but powerful ice-breakers are needed to keep parts of this route open.

The exploration of the Northwest Passage through the Canadian Arctic islands has a much more chequered history. Although the search for navigable routes to the Far East was of prime importance, the prospect of mineral wealth was also a compelling objective. An early attempt was made by Sir Martin Frobisher in 1576. This expedition was poorly led and produced minimal return. Frobisher even brought back pyrites or fool's gold from southern Baffin Island but, of course, this was of no commercial value. The seventeenth century saw several subsequent attempts, but all were unsuccessful and occasionally disastrous. However, there was one commercial outcome – the founding of the Hudson's Bay Company in 1670 for developing a fur trade with the indigenous peoples. The company still operates today.

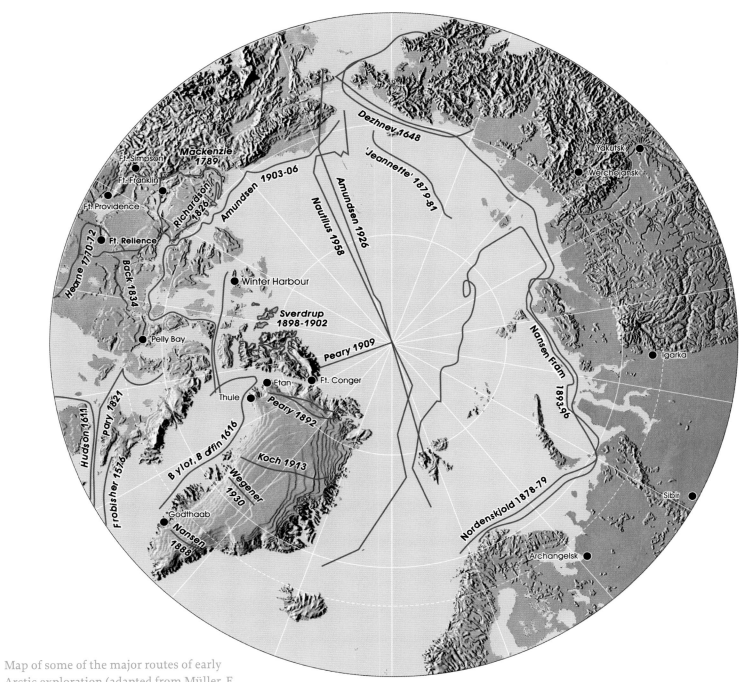

Map of some of the major routes of early
Arctic exploration (adapted from Müller, F.
1981. *The Living Arctic*. Methuen, London).

A Russian icebreaker uses its blunt bows to ride over the shorefast sea ice covering one of the deep sounds in the Franz Josef Land archipelago. The weight of the ship, the nuclear-powered 'Taymyr', then breaks the ice. The straight pathways cut through winter sea ice by icebreakers can even be seen as dark lines on satellite images of the Arctic taken from space over 150 km above the Earth's surface.

After all these failures, interest in a search for the Northwest Passage waned, but renewed attempts were made from 1818 onwards. The incentive for this was to maintain the pre-eminence of the British Navy following the Napoleonic Wars, and to continue employment for Navy personnel. At this time, two main expeditions were initiated: Captain David Buchan and (the then) Lieutenant John Franklin making up a western group, and Commander John Ross and Lieutenant Edward Parry a northern group. The Northwest Passage remained elusive, but the importance of Lancaster Sound as a key to the passage was recognised. At the same time the commercial prospects of whaling were established, soon leading to a vigorous industry. A number of privately-funded expeditions were also initiated. Foremost among them was John Ross with his nephew James Clark Ross, the latter destined to become one of the greatest Antarctic explorers. Their cruise lasted four years from 1829–1833 and included the first visit to the North Magnetic Pole by the younger Ross. Surveys were also made overland, often with help from the Hudson's Bay Company.

A major expedition was undertaken by the British Admiralty under the leadership of Sir John Franklin in 1845, with two ships, *Erebus* and *Terror*. The ships became trapped in the ice of Lancaster Sound. By the end of the second winter, Franklin had died. The remaining crew of 105 men later set off on a hopeless 'death' march across the sea ice in an attempt to reach safety on the mainland.

With the failure of Franklin and his crew to return, a series of forty relief parties was sent out, first to rescue any survivors and later to establish the expedition's fate. However, it was not until 1853 that Robert M'Clure's expedition first traversed the Northwest Passage (but by using sledge over sea ice rather than with his ship). The fate of Franklin's men had been determined by 1859, by the discovery of a trail of bodies. Only in 1903–1906 did the first ship sail through the Northwest Passage. Roald Amundsen of Norway was the successful navigator and, in addition, remapped the now much displaced North Magnetic Pole.

Attempts on the North Pole

The northernmost coasts of the Arctic islands fringe the Arctic Ocean, and have been the point of departure for both early and modern attempts to reach the North Pole. Once again, the initial objective was to find a shorter route to China or to discover a new landmass over the Pole. A series of expeditions from Britain, by Constantine John Phipps (1773), William Scoresby (1806), Buchan (1818) and Parry

The severe features of Roald Amundsen are depicted in this bust of the famous Norwegian explorer, at Ny-Ålesund.

(1827), all tried to force a way through the pack ice off Svalbard, but each was forced to retreat. In 1879, the American G.W. De Long made an attempt from the Bering Strait, but his ship, the *Jeanette*, after entering the pack ice near Herald Island, was crushed by the ice close to Bennett Island in 1881. Three years later, the finding of items from the *Jeanette* on drifting sea ice off the southwest coast of Greenland indicated the westward direction of ice drift across the Arctic Ocean for the first time.

Using this new oceanographic information, Fridtjof Nansen of Norway attempted to reach the Pole by allowing his ship to be frozen into the drifting pack ice. For this purpose, he commissioned *Fram* with a rounded hull that would be forced upwards when subjected to pressure from the ice (the ship is now preserved in a museum in Oslo). He sailed east in 1894 across the Barents and Kara seas. Off the New Siberian Islands, the *Fram* became icebound and, over the next three years, drifted north and west across the Arctic Ocean and reached a latitude of 85°57′ N. When it became apparent that the ship would not drift nearer to the Pole, Nansen and Hjalmar Johansen took a dog-sledge party north in an attempt to reach the Pole, but in so doing forsook the chance of returning to civilisation in *Fram*. They reached a latitude of 86°14′ N, before having to retreat south and reach Franz Josef Land, where they were transported home after a fortuitous meeting with Frederick Jackson, who was leading an Arctic expedition in his ship *Windward*. Meanwhile, the *Fram*, under the control of his second-in-command Otto Sverdrup, continued its drift and was finally released by the ice off the coast of Svalbard. Although Nansen did not reach his prime objective, the expedition proved to be a model of organisation, and achieved much of scientific value, especially in the fields of oceanography, marine biology and meteorology. It demonstrated, for example, that the central Arctic was a deep ocean. In later life, Nansen became well known for his humanitarian work with refugees following World War I, and was awarded the Nobel Peace Prize.

A further attempt on the North Pole was by the American, Frederick Cook, who in 1908 claimed to have reached the Pole from Axel Heiberg Island. However, his credibility was undermined by his claim, later proved false, of having reached the top of North America's highest peak, Mt McKinley (Denali), and by purporting to show a picture of the Pole that had rock visible in one corner! Cook was followed closely by another American, Robert Peary, who persuasively claimed to have reached the Pole in 1909 from Ellesmere Island. For many years, this claim was widely accepted and Peary gathered many influential supporters. However, many now doubt that he could have made the exceptionally fast daily journey times (50 km a day) that he claimed across fractured sea ice. If the claim of Peary was in fact false, then the first people to stand at the North Pole were 24 members of a secret Soviet airborne expedition in 1948. The first traverse across the whole Arctic Ocean from Alaska to Svalbard was Sir Wally Herbert's dog-sledging party from Britain in 1969.

In the late nineteenth and early twentieth centuries, a number of airborne attempts to reach the North Pole were also initiated. Salomon August Andrée's balloon expedition in 1897 ended in tragedy, with the three participants dying on the isolated White Island (Kvitøya), east of Spitsbergen. In 1926 the airship *Norge*, with Roald Amundsen and the Italian Umberto Nobile aboard, travelled across the Pole in an epic 5,100 km flight from Norway to Alaska. Nobile's attempt at a second overflight, in the airship *Italia* in 1928, ended in disaster as the airship crashed during its return journey to Spitsbergen. Amundsen, leader of the first expedition to reach the South Pole, subsequently disappeared over the Barents Sea during the rescue operation set in motion after the *Italia* crashed.

The mountains of Brøggerhalvøya form a backdrop for the airship mast at Ny-Ålesund, Spitsbergen, where Amundsen and Nobile started their journeys to the North Pole in the early years of the twentieth century.

The North Pole remains as difficult as ever to reach unsupported, and even with modern advances in equipment, many expeditions fail. However, nowadays a number of tourist groups depart each spring from Eureka on Ellesmere Island in ski-equipped Canadian aircraft, or from Sredniy in Severnaya Zemlya in large Russian helicopters. These aircraft refuel from caches *en route*, and land on the sea ice which constantly drifts across the vicinity of the Pole. Even larger numbers of tourists can join cruises to the Pole on giant nuclear-powered Russian icebreakers.

Scientific research

A shift of emphasis in polar activity from exploitation of marine resources and establishing trade routes was demonstrated by Nansen's *Fram* expedition. However, widespread systematic scientific investigation of the geology, glaciology and biology of the Arctic islands began only after World War One. Geological investigations by Lauge Koch of Denmark in North Greenland in 1920–23 provided a foun-

A Russian jet refuels at the air strip on Sredniy Island, Severnaya Zemlya. The runway and apron have a surface of compacted snow and ice. Each year the strip is only available before spring melting reduces the strength of its surface and makes it unusable for landing and take-off. Tourists and expeditions making for the North Pole usually arrive in April, to be transported northwards by helicopter.

Danish geological expeditions, led by Lauge Koch, mapped large tracts of East and North Greenland in the 1950s. Their central base, named 'Eagle's Nest', occupies an idyllic position on Ella Ø, a beautiful island in Kong Oscar Fjord.

dation for a new series of expeditions to East Greenland in the 1950s. These expeditions were manned by strong international teams of geologists and supported by ship and aircraft. Vast areas of rugged mountain country were mapped both geologically and topographically. The pioneering investigations of Hans Ahlmann of Sweden on the glaciers of Svalbard in the 1930s, and the many scientific experiments relating to the glaciology, geology, meteorology and the upper atmosphere undertaken in the Arctic islands as part of the International Geophysical Year in 1957–58, laid the foundations for research that continues in the Arctic today. Modern research in the Arctic islands includes not only continuing mapping by government surveys, such as the Geological Survey of Canada, the Geological Survey of Denmark and Greenland, and the Norwegian Polar Research Institute, but also focuses on resolving major scientific questions concerning, for example, global environmental change, in which universities play a leading role.

The Arctic remains a challenging environment for undertaking scientific research, although the infrastructure now in place has made even the remotest places accessible if sufficient finance is available. In the Canadian Arctic, much scientific fieldwork is supported logistically by the Polar Continental Shelf Project, which charters light aircraft such as de Havilland Twin Otters capable of landing on sea ice, lake ice or glaciers using skis, or on the tundra, raised beaches or gravel plains using balloon tyres.

In Greenland, research is co-ordinated by the Danish Polar Centre in Copenhagen, and they offer advice and information for scientific groups intending to undertake fieldwork there. Transport has to be arranged privately, but boats, helicopters and light aircraft are all available. Special permission is required to work in the Northeast Greenland National Park which covers nearly 1 million square kilometres.

In Svalbard, the Governor or Sysselmann requires all field parties to register their expeditions, whilst the Norwegian Polar Institute compiles a summary of all

A large Russian helicopter lands on Hooker Island in Franz Josef Land. This was the site of several scientific expeditions to the archipelago, including a major study of the nearby ice cap during the International Geophysical Year, 1957–58. The Russian cross is a memorial to the Sedov expedition of 1913–14. Georgiy Sedov died in 1914 after setting out for the North Pole from Franz Josef Land.

Freighting of supplies for fieldwork, especially when dependent on air support, is expensive. Careful planning of meals to maximise calorific value and minimise weight is therefore essential. Here, the contents of a Cambridge University 12 person/day food box, dominated by dehydrated foods, are laid out to illustrate the basic meal constituents on a self-contained expedition.

Teams of researchers often need to camp on Arctic islands for extended periods to make scientific measurements. Here, a tent camp has been established on the surface of the large Austfonna ice cap in Nordaustlandet, eastern Svalbard. Personal survival techniques are an important skill for the polar researcher. The scientists are part of a joint expedition between the Norwegian Polar Institute and the Scott Polar Research Institute.

scientific activities each year for general circulation and to facilitate collaboration. Travel to the main settlement, Longyearbyen, is easy and reliable by scheduled flight, and there are regular internal flights to the scientific station at Ny-Ålesund and the coal mining towns at Svea and Barentsburg. Away from these settlements, a variety of modes of transport is possible. Motorboats provide access to many of the sites around the western and northern coasts of Spitsbergen in summer, although sea ice can occasionally present problems. Alternatively, charter of helicopters is feasible, but expensive, and often unreliable due to the weather conditions. The interior icefields can be reached by snowmobile or other tracked vehicles. This is easy in spring, but in summer melt streams and snow swamps can make access difficult. Almost 60% of Svalbard is protected, and special permission is required to visit these areas; indeed some regions are out-of-bounds, especially during the breeding season of nesting birds, seals and walrus.

The scientific research base operated by the Canadian Polar Continental Shelf Project at Resolute Bay on Cornwallis Island at the northern shore of Lancaster Sound. The base is the communications hub for most scientific research expeditions to the Queen Elizabeth Islands. In the Canadian Arctic, English, French and the Inuktitut language of the Inuit are all used.

Many scientific groups in the Canadian Arctic are serviced by versatile short take-off and landing aircraft like this de Havilland Twin Otter. Such aircraft can be fitted with skis or balloon tyres for landing on snow or rough ground. The two turbo-prop engines are also mounted relatively high above the ground to avoid damage from stones when landing on dirt airstrips. This aircraft has just been refuelled at the airstrip close to the Canadian weather station at Eureka in northern Ellesmere Island.

The Russian settlement of Barentsburg in Spitsbergen. The mural celebrates the coal miners of the town.

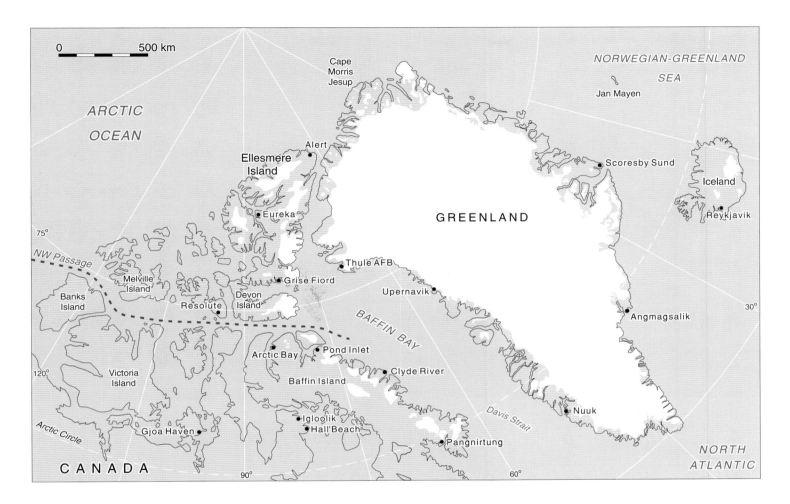

0 500 km

Cape Morris Jesup

NORWEGIAN-GREENLAND SEA

Jan Mayen

ARCTIC OCEAN

Alert

Ellesmere Island

Scoresby Sund

Iceland

Eureka

GREENLAND

Reykjavik

75°

NW Passage

Thule AFB

Melville Island

Grise Fiord

Upernavik

30°

Banks Island

Devon Island

Resolute

BAFFIN BAY

Angmagsalik

Arctic Bay

Pond Inlet

120°

Victoria Island

Clyde River

Baffin Island

Nuuk

Davis Strait

Arctic Circle

Igloolik

Gjoa Haven

Hall Beach

Pangnirtung

NORTH ATLANTIC

CANADA

90°

60°

Map of the eastern Canadian Arctic and Greenland with the main settlements located and the Northwest Passage indicated. Glaciers are indicated in white.

Place names in the Arctic islands

Place names give a clue to the activities of the first explorers of the Arctic islands and surrounding seas. In different parts of the Canadian Arctic, for example, M'Clure Strait and McClintock Inlet were named after British Navy officers from the Franklin search era; the names of American military men Kane and Greely were given to Kane Basin and Greely Fiord; Axel Heiberg and Amund Ringnes islands were named by the Norwegian Sverdrup expedition of 1899–1902. On Ellsemere Island, Judge Daly Promontory and Clements Markham Inlet were named after former presidents of the American Geographical Society and the Royal Geographical Society, respectively. These organisations, along with a somewhat less

Small motor boats have proved an effective way of undertaking geological investigations along the indented coastline of Spitsbergen in summer. Here, two vessels, the *Collenia* (left) and the *Salterella* (right), operated by the Cambridge Spitsbergen Expeditions through the 1970s and 1980s, are anchored in front of the glacier Scottbreen in Bellsund. These expeditions, which were led by Brian Harland (right), undertook geological mapping in Svalbard.

Snowmobile transport in spring is fast and efficient, but extreme care is needed especially on sea ice, as depicted here around a debris-bearing iceberg in Kongsfjorden, northwestern Spitsbergen.

In good weather, helicopters are the most efficient means of transport to a field site. Here we see a Bell 212 helicopter off-loading a geological party near Sorteelv Gletscher in East Greenland. The geologist is stooping by the equipment to prevent items blowing away under the draught from the rotor blades as the helicopter takes off.

likely Danish brewing company, after which Tuborg Lake is named, sponsored expeditions to the Canadian Arctic islands.

In the Eurasian north, Svalbard also bears the names of its explorers. Some names with a British origin, for example on the isolated eastern island of Nordaustlandet, derive from Oxford and Cambridge University expeditions. Many features in northeast Spitsbergen are named after Cambridge colleges and a glacier was even named after a bakery, Fitzbillies in the same city! The Austrian explorers of the archipelago known as Franz Josef Land, named the islands after their Arch-Duke. Ostrov Bol'shoy Lyakhovskiy and Ostrov Malyy Lyakhovskiy, in the New Siberian Islands, are both named after the eighteenth century Yakutsk trader and trapper Ivan Lyakhov, who first reported them. A larger and more northerly island, Ostrov Kotel'nyy, is named after a copper kettle or 'kotel' found there by Lyakhov! The small islands of the De Long archipelago, Jeanette, Bennett and Henrietta islands, were named after the ship in De Long's expedition, the expedition sponsor

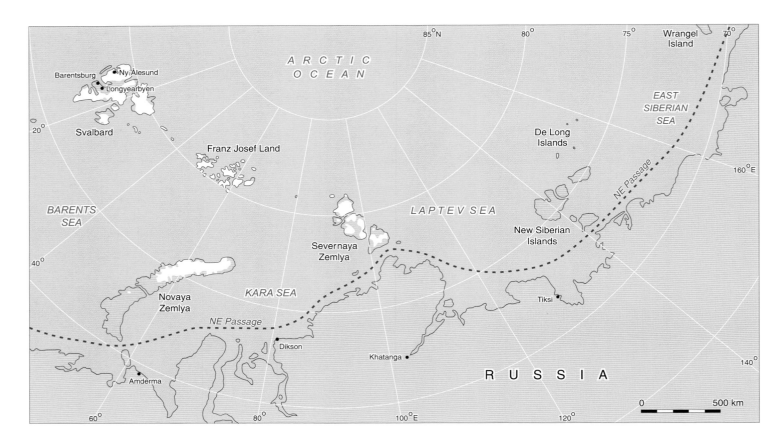

Map of the Eurasian Arctic with the main settlements located and the Northeast Passage or Northern Sea Route indicated. Glaciers are indicated in white.

and his mother, respectively. Severnaya Zemlya was the last major land mass to be discovered anywhere on Earth. It was first sighted in 1913, but not mapped until 1930–32.

Strategic importance and politics in the Arctic islands

From World War II onwards, the High Arctic acquired considerable strategic significance. Aircraft produced in North America were ferried to Britain via a number of isolated airstrips in Northern Canada and Greenland. Some still lie where they crashed *en route*, now buried by the snows of the Greenland Ice Sheet. Atlantic convoys of ships, bringing food and fuel to Britain and to Northern Russia, also sailed through Arctic waters, for example the Denmark Strait between East Greenland and Iceland.

The writing on a 45-gallon drum of jet fuel testifies to a strong military presence in the Arctic. The fuel is consigned to the Commander of the 12th Missile Warning Group at Thule Air Force Base in Northwest Greenland. Note the mis-spelling 'Artic' on the drum!

During the 1950s and 1960s, at the height of the Cold War, the shortest route for conventional bombers and intercontinental ballistic missiles between the Soviet Union and the United States was across the Arctic Ocean. The Arctic islands became the frontier in the silent tracking of nuclear submarines beneath the Arctic sea ice, and for constantly watchful radar systems to monitor the Arctic skies. A series of 42 Defense Early Warning or DEW-Line radar stations was built by the USA to straddle the Arctic; locations included Alert at the northern tip of Ellesmere Island, Dye 1, 2 and 3 on the Greenland Ice Sheet, and Ny-Ålesund in Spitsbergen. Large airstrips, acting as advanced bomber and reconnaissance aircraft bases, were constructed by the Americans at Thule in Northwest Greenland and by the Soviets in Franz Josef Land and Wrangel Island. Thankfully, many of these installations are now of no more than historical significance, as Cold War tensions have eased in the wake of political change in Russia. One of the legacies of the Cold War is, however, the presence of very large numbers of aging 45-gallon fuel drums, often located in isolated caches, to provide emergency supplies. Leakage is a potentially serious environmental hazard.

More recently, the Arctic Council was established in 1996, with representatives coming from all the Arctic-rim nations, plus four indigenous people's

Abandoned and rusting fuel drums, either empty or still containing potentially hazardous liquid, are a common sight around many settlements in the Arctic islands. These are at the Russian base on Sredniy Island in northwest Severnaya Zemlya. Snow still covers much of the ground in the Arctic spring, but, when summer snow disappears, the full extent of environmental damage is revealed all too often.

organisations, and observers from those other countries with well-established Arctic interests. The Council provides a high-level diplomatic setting in which discussions and policy-formulation can take place between member states. Despite certain sovereignty issues, for example the status of the Northwest Passage as an international waterway, the Arctic at the beginning of the twenty-first century is no longer a frontier region for military activity. Instead, both its landscape and the native peoples who have lived in the Arctic islands for millennia are recognised increasingly as important parts of the physical and cultural richness of our Planet.

Mineral exploration

Exploration for, and exploitation of, natural resources such as precious and semi-precious metals, oil, gas, coal and cryolite has also gone on throughout the twenti-

The remains of an ill-fated early twentieth century attempt at marble extraction by the Northern Exploration Company on Blomstrandøya, northwestern Spitsbergen, showing horizontal and vertical steam boilers. The settlement established was known as 'London', but was active for barely more than one season, as the marble disintegrated when the permafrost melted.

eth century. In the early years of the century, there was strong interest in mineral exploration in Spitsbergen, much of it undertaken by the Scottish Spitsbergen Syndicate, whose derelict mining camps may still be seen. In 1920, the Spitsbergen Treaty was signed, and in 1925 sovereignty was conferred on Norway, making specific allowance that other signatory nations could continue to exploit minerals in the archipelago. Even short lengths of railway track were constructed as far as 80 °N in association with coal-mining. The mining of coal by both Norwegian and Russian companies continues today.

Mining for high-value minerals takes place in many parts of the Arctic. In Greenland, extraction of cryolite, an aluminium ore, began at Ivittuut as early as 1858, but since then only a few mines have been active, such as a lead-zinc mine near Uummannaq in West Greenland. However, in recent years exploration for diamonds and gold has been undertaken, and it is anticipated that new mines will be opened soon. The Greenland authorities encourage the local people to assist in the

A repainted, but non-operational, steam locomotive with several coal wagons, testifies to the once thriving, but now defunct, coal mine at Ny-Ålesund. Today, with the scenic backdrop provided by the glacier Conwaybreen, the train provides a favourite photographic opportunity for the many cruise-ship visitors to the settlement.

search for minerals, by asking them to send unusual rock samples for laboratory analysis, and offering a prize for the most promising specimen.

Mines for scarce minerals also operate throughout the Canadian Arctic islands, including the lead-zinc mines of Polaris on Little Bathurst Island and Nanisivik in northern Baffin Island. These installations employ up to several hundred people at each site. There was once a gold mine on Bolshevik Island in Severnaya Zemlya.

The search for hydrocarbons in the Barents Sea region has been characterised by short bursts of intense activity. As a result, oil and gas have been found in the sedimentary basin of central Spitsbergen, although not in sufficient quantities to be commercially viable. By contrast, the southeastern Barents Sea, close to the southern tip of Novaya Zemlya, has some of the largest natural gas reserves in the World. Hydrocarbon exploration around Greenland began in the mid-1970s when five wells were drilled in sedimentary basins offshore of West Greenland. The search was resumed in the late 1990s in this area. The East Greenland region of Jameson Land received some exploration activity in the 1980s, but no wells were drilled, yet geologically the potential remains, as equivalent, once adjacent, rocks around the north British continental shelf have proved to be productive.

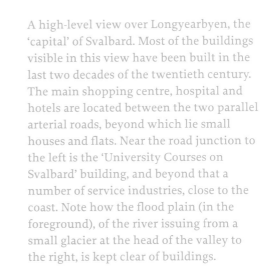

A high-level view over Longyearbyen, the 'capital' of Svalbard. Most of the buildings visible in this view have been built in the last two decades of the twentieth century. The main shopping centre, hospital and hotels are located between the two parallel arterial roads, beyond which lie small houses and flats. Near the road junction to the left is the 'University Courses on Svalbard' building, and beyond that a number of service industries, close to the coast. Note how the flood plain (in the foreground), of the river issuing from a small glacier at the head of the valley to the right, is kept clear of buildings.

Tourism and education

The most recent source of growth in the Arctic islands has been stimulated by tourism. The Arctic is an attractive destination because of its landscapes and wildlife, and it is rightly perceived as one of the last truly unspoiled wilderness areas on the globe. The character of settlements within the Arctic islands is changing as a result of this influx. Indeed, tourism is becoming a progressively more important part of the northern economy as the strategic value of the Arctic declines. Longyearbyen in Svalbard, for example, was a town that grew up following the opening of the first coal mine by John M. Longyear in 1906. The mining activity was taken over by the Store Norsk Spitsbergen Kullcompani ten years later, and until the mid-1980s Longyearbyen remained very much a company town. During the past decade, however, new hotels and tourist companies have been established, along with a full infrastructure including schools and the University Courses on Svalbard (UNIS), a branch of the Norwegian university system that reaches out to the international community. Longyearbyen is the seat of local government and now has a resident population of about 1,400 people. Similarly, the

The Arctic is increasingly seen as a place where families live. In Svalbard, the community of Longyearbyen has been transformed from a mining settlement to a fully-integrated community over the past fifteen years or so. The church is one of the older community buildings, located near the cliffs which dominate the settlement.

Ny-Ålesund is a former coal-mining village, but closed down following a major disaster in 1962. The Norwegians have maintained a scientific station there for many years. Within the last decade, the settlement has found a new lease of life, with the establishment of scientific stations by several nations. The old buildings have been renovated, and new ones, with purpose-designed laboratories, have been built. The view across the fjord to the conical peaks of the Tre Kroner, or Three Crowns, and Dronningfjella is a feature of the settlement's beautiful location.

former coal-mining village of Ny-Ålesund is now run as an international research centre, with Norway, France, Germany, Italy, Japan and the UK all operating stations there. In 1999 it was estimated that there were about 35,000 visitors to Svalbard, the majority arriving on cruise ships.

Care is being taken to see that the growing tourist activity does not result in damage to the very environment that has stimulated it. In Ellesmere and Baffin islands in Canada, in Svalbard, and even in Russian Franz Josef Land, national parks and protected areas have been set up, with specific rules governing, for example, the disturbance of colonies of nesting birds by overflying aircraft. The operators of tourist ships visiting Moffen, north of Spitsbergen, where a large walrus colony is located, have to obey strict rules concerning landing. Archaeological sites have also been mapped and are protected by law. These sites range from ancient camps of the Dorset culture and the remains of Franklin's winter quarters at Beechey Island in the Canadian North, through to whalers' graves and even German World War II meteorological stations in Svalbard.

Tourism is an increasingly important part of the economy of the Arctic islands. Cruise liners bring large numbers of tourists to visit the fjords and settlements of Spitsbergen. The settlement of Ny-Ålesund (right) is dwarfed by a cruise liner which carries over 1,500 passengers, while the fjord walls of Lilliehöökfjorden, an arm of Krossfjorden, tower above another tourist ship (above).

Chapter 10

Postscript: the future of the Arctic islands

The physical environment of the Arctic islands is likely to be affected in the foreseeable future by the World's changing climate, just as it has been at longer timescales by past changes in global climate. Sophisticated computer models of our atmosphere and oceans predict that temperatures will rise more in the Arctic than anywhere else on the globe in response to increases in 'greenhouse' gases and other atmospheric pollutants. A warming of between 2 and 8 °C is predicted over the next 100 years, although these computed values are estimates averaged over very large areas. Temperatures will continue to vary from year to year and from place to place within the Arctic, although with a likely upward trend. Precipitation too is likely to rise. Records of past climate, preserved in long ice cores, also show that the environment of the Arctic islands has been variable over the last 100,000 years or so.

These environmental changes will affect both the physical and the biological systems of the Arctic islands. Implications of warming for plants and animals include a longer summer season for plant growth and for migratory breeding birds. Permafrost may melt to a greater depth during warmer summers, increasing soil moisture, the area of Arctic wetlands and the hazards to building foundations. The melting of previously frozen ground also releases methane, a greenhouse gas, to the atmosphere. Any changes in ocean temperature will also affect the distribution of marine organisms.

The vast ice sheets covering Greenland and Antarctica will probably not thin or retreat greatly over the next century or so, largely because the increased melting at their margins is likely to be offset by a rise in snowfall in their interiors as

Brief March sunshine illuminates the jagged mountains of East Greenland which project above outlet glaciers of the Greenland Ice Sheet.

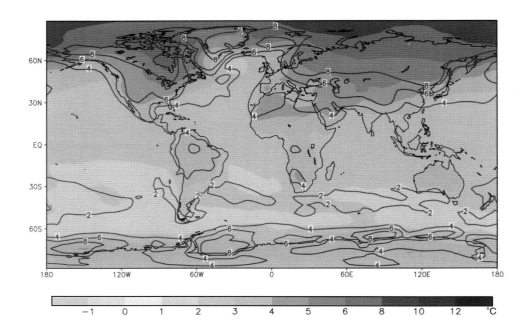

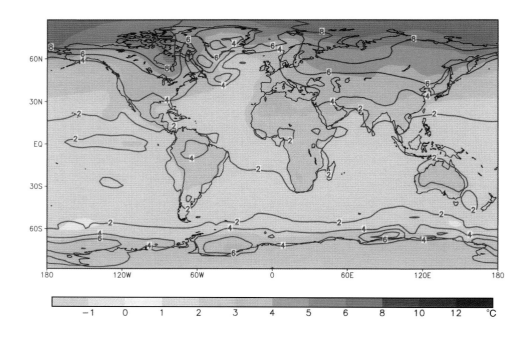

Two predictions of climate warming from Atmospheric General Circulation Models (GCMs). Note that, in both cases, mean annual temperatures are likely to increase more in the Arctic than elsewhere on the planet. (Adapted from the IPCC, 2001. *Climate Change 2001: The Scientific Basis.* Cambridge University Press.)

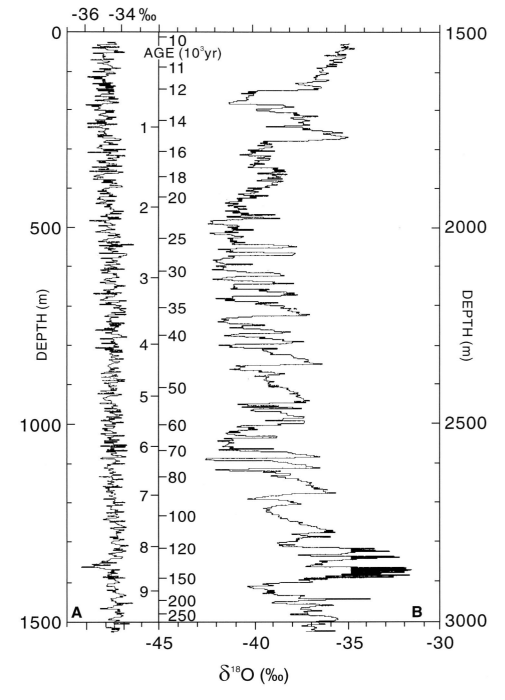

The record of climate change over the past 100,000 years, derived from a 3,000 m-long ice core through the Greenland Ice Sheet. Oxygen isotope values ($\delta^{18}O$) are measured on ice samples from the core, and their fluctuations are a reflection of past changes in temperature and the amount of ice on land. The changes from coldest (most negative isotopic values) to warmest represent a temperature change of between 15 and 20 °C. The left-hand panel shows values for the past 10,000 years, and the right-hand panel those over the past 100,000 years or so. Ice older than about 100,000 years may have been disturbed by flow, and a climate record cannot be interpreted at greater depths in the core. (Adapted from W. Dansgaard *et al.*, 1993. *Nature*, vol. 364, pp. 218–220.)

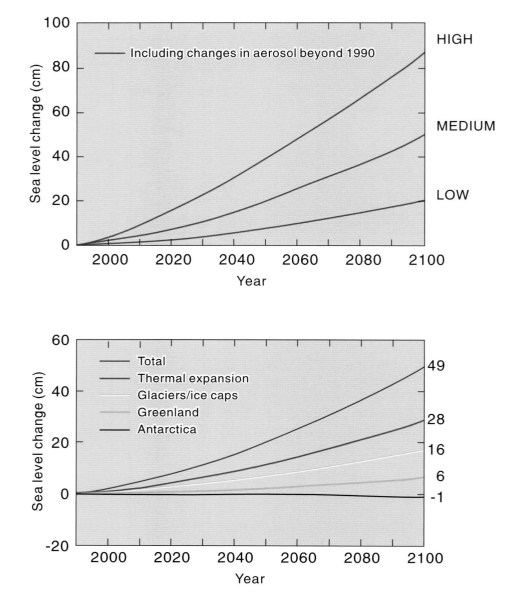

Projections of sea-level rise over the next 100 years in a warming World. The upper diagram shows an envelope of estimates of possible rise, with 50 cm being the most probable value. The lower diagram indicates the likely contributions to sea level from glaciers and ice caps (including those of the Arctic), and the Greenland and Antarctic ice sheets. Thermal expansion takes place as the oceans of the World warm. (Adapted from IPCC, 1996. *Climate Change 1995: The Science of Climate Change.* Cambridge University Press.)

temperatures rise at high latitudes. By contrast, the smaller ice caps and glaciers of the Arctic, together with those in mountainous areas such as the North American Cordillera, South American Patagonia, the European Alps and the Himalayas, are much more sensitive to shifts in climate. Increased surface melting, resulting in glacier thinning and retreat is expected. This meltwater flows back into the oceans of the World and will result in a sea-level rise that is predicted to be between 20 and 90 cm over the next 100 years. This will affect the coastline of the Arctic islands, although perhaps less than the major port cities of Europe and North America. This is because the Arctic islands are still experiencing slow crustal rebound from the removal of past ice loading, which is recorded in the raised beaches illustrated in Chapter 7. This uplift is itself a result of past climate changes which caused the growth and subsequent decay of large ice sheets over much of the North American and Eurasian Arctic.

In addition, both the ice floes making up polar pack ice and the coastal shore-fast sea ice are predicted to become thinner and less extensive. As well as having the positive feedback effect towards further warming discussed in Chapter 5, both the activities of wildlife and human travel over the sea ice will be affected. For communities within the Arctic islands, spring travel over sea ice by snowmobile will

Sea-ice floes are found throughout the Arctic seas, as in this example of regularly-shaped floes in an Ellesmere Island fjord.

The Sun's last rays illuminate the entrance to Scoresby Sund in East Greenland.

become more difficult, whereas summer access to Arctic settlements by ship will be easier, with possible commercial development of cargo transport routes through the Northwest Passage and north of Siberia. For seal, walrus and polar bear populations, changes in sea-ice conditions will reduce the habitat in which they hunt and breed, with related effects on native hunters.

If the prediction of climatic change in the islands of the Arctic is not without its assumptions and difficulties, the future of human activity in the Arctic is yet more problematic. During the past few hundred years or so the Arctic islands have first been explored, and later mapped and to some extent exploited for their biological and mineral wealth. The northern native peoples have gone from a traditional way of life in balance with the very harsh natural environment in which they live, through initial interactions with lower-latitude peoples interested in commercial and strategic gain and also with issues of sovereignty. Most recently, the increasing recognition of the rights of native peoples, and of the significance of their culture and traditional way of life, has led to the establishment of Nunavut in Canada and home rule for Greenland.

Change is likely to continue in the Arctic islands. Military interest in the Arctic has declined since the end of the Cold War, and tourism appears likely to become an increasingly important component of northern activity. Tourists come to enjoy the landscape and wildlife of the Arctic islands, and it is widely recognised that visitors and organisers alike should strive to minimise any environmental damage and to manage access to sensitive areas. It is our hope that this book will enhance the enjoyment of those who come to the islands of the Arctic, and to awaken the interest of those who have yet to appreciate its unique landscapes, wildlife and peoples.

The rapid recession of tidewater glaciers is
sometimes linked to climatic warming.
Here, recently calved icebergs derived from
Kronebreen litter the shore of
Kongsfjorden in northwest Spitsbergen.

Suggested further reading

Barr, S. (Ed.) 1995. *Franz Josef Land*. Norsk Polarinstitutt, Oslo. [comprehensive pocket guide for non-specialists]

Escher, A. & Watt, W. S. 1976. *Geology of Greenland*. Grønlands Geologiske Undersøgelse, Copenhagen. [superbly illustrated definitive summary of Greenland's geology; specialist knowledge required]

Feilberg, J., Fredskild, B. & Holt, S. 1984. *Grønlands Blomster/Flowers of Greenland*. Forlaget Regnbuen, Ringsted, Denmark. [well-illustrated pocket guide]

Gjærevoll, O. & Rønning, O.I. 1980. *Flowers of Svalbard*. Universitetsforlaget, Oslo. [well-illustrated pocket guide]

Hambrey, M. J. & Alean, J. 1992. *Glaciers*. Cambridge University Press, Cambridge. [well-illustrated explanation of glaciers and their landscapes for the lay person]

Hjelle, A. 1993. *Geology of Svalbard*. Norsk Polarinstitutt, Oslo. [pocket handbook for non-specialist]

Mehlum, F. 1990. *Birds and Mammals of Svalbard*. Norsk Polarinstitutt, Oslo. [well-illustrated pocket guide]

Müller, F. 1977. *The Living Arctic*. Atlantis Verlag AG, Zurich.

Nuttall, M. & Callaghan, T. V. (Eds.) 2000. *The Arctic: Environment, People, Policy*. Harwood Academic Press, Amsterdam. [wide-ranging academic reviews of physical, human and strategic aspects of the Arctic]

Sage, B. 1986. *The Arctic and its Wildlife*. Croom Helm, London. [scholarly work on Arctic flora and fauna, and their adaptation to the environment]

Stonehouse, B. 1971. *Animals of the Arctic: the Ecology of the Far North*. Eurobook Ltd/Peter Lowe, Ipswich, England [comprehensive, well-illustrated review]

Sugden, D. 1982. *Arctic and Antarctic: A Modern Geographical Synthesis*. Blackwell, Oxford. [a bipolar comparison for geography and environmental science undergraduates]

Vaughan, R. 1999. *The Arctic: A History*. Sutton Publishing, Stroud, England. [comprehensive, highly readable history]

Worsley, D. & Aga, O.J. 1986. *The Geological History of Svalbard*. Den Norske Stats Oljeselskap a.s. (Statoil), Stavanger, Norway. [well-illustrated review with colour photographs; some specialist knowledge required]

Geographical index

Numbers in bold refer to illustrations. Those in italics refer to maps.

Subject index

Numbers in bold refer to illustrations.